The Martin Gardner Bibliography

The
Martin Gardner Bibliography

೫

by
Dana Richards

Foreword by
Donald E. Knuth

CSLI PUBLICATIONS
STANFORD, CALIFORNIA

Names: Richards, Dana, 1955– editor.

Title: The Martin Gardner Bibliography, Foreword by Donald Knuth /
by Dana Richards.

Description: [Stanford, CA] : CSLI Publications, [2023] | Includes bibliographical
references and index. | Summary: "Martin Gardner was a polymath whose
international reputation extended from mathematics to literature, from
philosophy to science, and from magic to fiction. He was the father of the
modern skeptical movement. This comprehensive bibliography covers every
aspect of Martin Gardner's lengthy publishing career, from 1930 to 2010, and
includes detailed descriptions and indexes of his writings on mathematics and
many other topics. Over two hundred boxes of Martin Gardner's mathematical
papers have until now existed only in the Stanford archives, and this
bibliography puts them all into perspective. Everything in this book is being
published for the first time. Dana Richards worked directly with Martin
Gardner from 1978 to 2010 on this project, and continues to access files in his
family's possession. On a book dedication page Gardner described him as a
"bibliographer extraordinaire". Dana Richards is a professor of Computer
Science at George Mason University and has previously edited two Gardner
books: The Collosal Book of Short Puzzles and Problems and Dear Martin/Dear
Marcello: Gardner and Truzzi on Skepticism. This book includes a foreword by
Donald Knuth, who dedicated his book Selected Papers on Fun and Games to
Martin Gardner"– Provided by publisher.

Identifiers: LCCN 2023005997 (print) | LCCN 2023005998 (ebook) |
ISBN 9781684000753 (ebook) | ISBN 9781684000623 (paperback)

Classification:
LCC Z8324.175 QA29.G268 (ebook) | LCC Z8324.175 .R53 2023 QA29.G268 (print) |
DDC 016.79374092 23/eng/20230–dc01

LC record available at https://lccn.loc.gov/2023005997

CIP

CSLI Publications is located on the campus of Stanford University.

Visit our web site at
http://cslipublications.stanford.edu/
for comments on this and other titles, as well as for changes
and corrections by the author and publisher.

Contents

IV Magic   235

19 Magic for the Public   239

20 Mathematical Tricks   245

21 Physics Tricks   249

22 Magic Literature   257

Foreword

I love to witness a Labor of Love! Dana Richards has spent more than half of his life accomplishing an almost impossible task: to catalog and put in order the writings of one of the most prolific authors the world has ever known.

Martin Gardner was an extraordinarily gifted writer, whose work has profoundly influenced my own life and that of thousands of others. The annual "Gatherings for Gardner," which began in 1993 and continue to grow today with frequent online webinars as well as meetings in person, make it clear that his works will continue to inspire new generations of readers for many years to come. Martin's delightful prose — and poetry! — has astonishing breadth and depth and volume. Now for the first time we can see it in its totality.

I was privileged to have known Martin since the 1960s, and to have corresponded with him frequently for more than 40 years. In 1994 I spent an unforgettable two weeks in Hendersonville, North Carolina, where I literally lived with the hundreds of files on recreational mathematics that he subsequently donated to Stanford, and with thousands of his books. (He and his wife Charlotte had purchased an extra home in which to store them.) Yet when I look at a random page of this bibliography, chances are good that I'll learn something that I didn't know about his enormous range of activities, about unexpected nuggets of wisdom that remain for me to explore.

Of special interest is Dana's brand-new index to The Canon, by which of course I mean Martin's decades of columns written for *Scientific American* magazine, later published as a classic series of books. I keep those fifteen volumes next to the chair where I work every day; and for 30 years I've been using photocopied indexes compiled by Carl Lee and by David Singmaster (with handwritten notes of my own) to find my way around. And when I compare the index in the present book to that oft-thumbed resource, I certainly like Dana's version much better. (For example, we can now readily see where Martin mentioned *MAD* magazine.) Dana has even included a magnificent index to each of the individual problems that appear within the chapters — like the problem of the absent-minded teller, or the problem of the rotating table. There are 250 of those, each with a one-line summary, plus 72 "quickies."

I got to know Martin because of his essays on mathematical games, but that topic accounts for only 30% of this bibliography. The remaining works can be roughly categorized as expositions of science, mostly physics (25%); of philosophy and literature (25%); and of magic (20%). Dana maintains the same level of quality in each of these categories. Indeed, he has succeeded heroically in ferreting out Martin's numerous contributions to periodicals about magic, which have always been intentionally obscure and difficult to locate.

In summary, this book sets a new standard of quality for bibliographies. It's an index that includes indexes of indexes. It gives details about essentially everything that Martin published, either under his own name or under a dozen aliases. It sheds important light on the day-by-day context in which a great thinker lived and moved, while communicating to the best of his ability the beauties and joys of our world. I shall certainly treasure this work on my bookshelves.

Donald E. Knuth

Introduction

Martin Gardner died in 2010, at the age of 95. Until the very end his mind and his writing were still sharp. In our specialized world I doubt we will ever again see someone who had an active involvement in so many intellectual spheres. He is well-known for his work in mathematics, science, philosophy of science, literature, theology, and magic. However few knew his full breadth.

After getting a B.A. in philosophy (University of Chicago, 1936) he attended a seminary for a year (to study philosophy). He realized that he was training himself for an academic position and that his real calling was to be an author. It was a decade before his writing allowed him to eat regularly. The expected career in fiction gave way to nonfiction, articles in science, philosophy and games.

He published something every year from 1930 (age 16) until his death, as well as several posthumous works. He was happiest with his fingers on the keys of a typewriter. He never wrote with a computer. He had a lot to say but never spoke in public, so there was a constant stream of writing. A lot of it is still unpublished, principally his correspondence, often quite detailed, that consumed hours most days.

The correspondence paid off when people started sending more and more his way, giving him even more material to write about. Everyone knew him and he knew everybody, it seemed. He was the generous conduit through which ideas flowed, in the magic, math and pseudo-science communities. His files started in shoeboxes while at college and grew to mammoth proportions. A description of using them, written in 1950s, sounds like a modern description of surfing the internet!

His interests were catholic. Often he found he was one of the first to write seriously about a subject. He trusted his instincts and helped create communities that flourish today. He never hid behind an academic demeanor. Gardner would be quick to remind us he was a "journalist." His topics could be erudite but he only had a lay reader in mind and he appealed to a wide audience.

The purpose of this book is two-fold. First, people in one of the communities he fostered are probably unaware of all of his contributions for that community. I have found this to be true many times. Second, most of his "fans" (almost an oxymoron for nonfiction) are largely unaware of his breadth. It is sometimes hard to convince them that all of these "Martin Gardner"s are the same. It is easy to be over-impressed by his output, but the truth is he was just well-organized (with *extensive* files) and disciplined. He was an incisive and clear thinker but a simple and humble man.

Details

This bibliography has been compiled over the last 45 years and it has some qualified claims to being complete. Martin Gardner was personally involved all that time, until his death in 2010, supplying copies of many of the items and pointing out others. (Often I would remind him of citations that had been missing/forgotten.) Of course, since new citations continue to appear it cannot be complete. To avoid repetitiveness, the British book citations appear with the "foreign" translations, unless they are the primary reference.

In each section the entries are in chronological order. However, I adopt the practice of listing variants, like a new publisher or an updated edition, with the original publication, so such variants will not appear in chronological order. Of course a new edition can be sufficiently different that they will be listed separately. Hence any attempt to count the number of books is problematic and it is not attempted. Gardner would often introduce corrections in different printings (as is evident in his correspondence with publishers) without any comment. No attempt has been made to identify these stealth variants.

Many of the citations that had been scattered over the years have appeared in a series of volumes of collected works. Very often there are updates and addenda in these collections. When an entry in one of these books is given, as an alternate citation, one of these abbreviations is used:

- *S:GBB — Science: Good, Bad, and Bogus*, 1981
- *O&S — Order & Surprise*, 1983
- *N-S P —The No-Sided Professor*, 1987
- *NA — The New Age: Notes of a Fringe-Watcher*, 1988
- *W&W — Gardner's Whys and Wherefores*, 1989
- *OTWS — On the Wild Side*, 1992
- *TNIL — The Night is Large*, 1996
- *WW&FL — Weird Water & Fuzzy Logic*, 1996
- *Adam&Eve — Did Adam and Eve have Navels?*, 2000
- *WJ — From the Wandering Jew to William F. Buckley Jr.*, 2000
- *GW — A Gardner's Workout*, 2001
- *UTB — Are Universes Thicker than Blackberries?*, 2003
- *JFH — The Jinn from Hyperspace*, 2008
- *FFGC — The Fantastic Fiction of Gilbert Chesterton*, 2008
- *T&F — When You Were a Tadpole and I Was a Fish*, 2009

In addition to these anthologies there are the collections of mathematical recreations columns that are detailed in the Mathematics Part.

Archives

Martin Gardner Papers (SC0647).
Department of Special Collections and University Archives,
Stanford University Libraries, Stanford, Calif.

Martin Gardner had more the twenty file cabinets of materials, both related to his writings and cuttings for use in the future. The files related to his mathematical writings, principally for every *Scientific American* column, had been sent to Stanford University, during his lifetime, in 2002. He did not need them at the time, but it became an issue after that when Cambridge University Press asked him to do updates. The files contain all the correspondence that was the lifeblood of the column. The collection is open to qualified researchers, but is not on-line. However the 75 boxes have been carefully indexed by Stan Isaacs and that index can be found at:

oac.cdlib.org/findaid/ark:/13030/kt6s20356s/entire_text/

Gardner's archives on pseudoscience are found at the headquarters of the Center for Inquiry (CSI) in Amherst, NY (formerly CSICOP).

Finally, any corrections / additions would be welcome, particularly for the Magic part and foreign editions. (richards@gmu.edu)

Part I

Mathematics

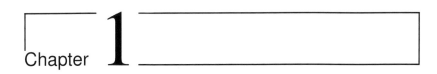

Chapter 1

Recreational Mathematics

Richard Guy said Martin Gardner "brought more mathematics to more millions than anyone else." Judging by public pronouncements, book dedications, thousands of letters written to Gardner, and general hearsay, no individual in history has influenced more people to enjoy and study mathematics. While he had little formal training in mathematics he was already familiar with recreational mathematics before the editors of *Scientific American* approached him to write a monthly column. He ended up writing nearly 300 columns.

His "Mathematical Games" column, starting in 1956, became so identified with the magazine that many readers reported it was the first thing they turned to. The editors acknowledged it was important to their sales figures.

While his mathematical training was initially weak, during his tenure ties with the mathematics community deepened, strengthening the rigor of the column. He always said that having to struggle with concepts helped him write clear explanations. Moreover he had developed a habit of finding connections from mathematical topics to the larger world, to literature, to art, to science, to philosophy and to magic. This was the secret of the column. He did not "make" mathematics interesting, he showed that it "was" interesting.

This topic arose later in his life but is presented first because users of this book expect it to be. However the long indexes to his column, which would disrupt the flow here, appear at the end.

1.1   Books

To avoid confusion, books of collected columns from *Scientific American* and *Isaac Asimov's Science Fiction Magazine* are listed here, even though they are discussed more fully in later chapters.

1. *Mathematics, Magic and Mystery* (Dover, 1956). Also cited in the Magic Part.

   - *Mathematics, Magic and Mystery* (BN Publishing, [2012]).

   Facsimile pirated edition

2. *The Scientific American Book of Mathematical Puzzles and Diversions* (Simon and Schuster, 1959). (Paperback 1963)

   - *Hexaflexagons & Other Mathematical Diversions: The First Scientific American Book of Puzzles & Games* (University of Chicago Press, 1988).

   "With a new afterword by the author, and an extensive new bibliography."

   - *Hexaflexagons, Probability, Paradoxes, and the Tower of Hanoi* (Cambridge University Press / Mathematical Association of America, 2008).

   A new preface by the author, and "much fresh material."

3. *The 2nd Scientific American Book of Mathematical Puzzles and Diversions* (Simon and Schuster, 1961). (Paperback in 1967)

   - *The 2nd Scientific American Book of Mathematical Puzzles and Diversions* (University of Chicago Press, 1987).

   "With a new postscript by the author."

   - *Origami, Eleusis, and the Soma Cube* (Cambridge University Press / Mathematical Association of America, 2008).

   Updated.

4. *New Mathematical Diversions from Scientific American* (Simon and Schuster, 1966). (Paperback in 1971)

   - *New Mathematical Diversions from Scientific American* (University of Chicago Press, 1983).

   "Phoenix edition."

   - *New Mathematical Diversions* (Mathematical Association of America, 1995).

   "With a postscript and a new bibliography."

- *Sphere Packing, Lewis Carroll and Reversi* (Mathematical Association of America / Cambridge University Press, 2009). Updated

5. *The Numerology of Dr. Matrix* (Simon and Schuster, 1967).
6. *The Unexpected Hanging and Other Mathematical Diversions* (Simon and Schuster, 1969).

   - *The Unexpected Hanging and Other Mathematical Diversions* (Simon and Schuster, 1969). (Fireside paperback.)
   - *The Unexpected Hanging and Other Mathematical Diversions* (Simon and Schuster, 1986). (Touchstone paperback.)
   - *The Unexpected Hanging and Other Mathematical Diversions* (University of Chicago Press, 1991).

     "With a new afterword and expanded bibliography."

   - *Knots and Borromean Rings, Rep-Tiles, and Eight Queens* (Cambridge University Press / Mathematical Association of America, 2014).

     Updated.

7. *Martin Gardner's Sixth Book of Mathematical Games from Scientific American* (W. H. Freeman, 1971).

   - *Martin Gardner's Sixth Book of Mathematical Games from Scientific American* (Charles Scribner's Sons, 1971).

     (Simultaneous paperback edition.)

   - *Martin Gardner's Sixth Book of Mathematical Diversions from Scientific American* (University of Chicago Press, 1983).

     "Phoenix edition" with slight title change.

8. *Mathematical Carnival* (Borzoi/Knopf, 1975).

   - *Mathematical Carnival* (Vintage, 1977).
   - *Mathematical Carnival* (Mathematical Association of America, 1989).

     Foreword by John Horton Conway, and "a new postscript and a new bibliography by Mr. Gardner."

9. *The Incredible Dr. Matrix* (Scribners, 1976).
10. *Mathematical Magic Show* (Borzoi/Knopf, 1977).

    - *Mathematical Magic Show* (Vintage, 1978).
    - *Mathematical Magic Show* (Mathematical Association of America, 1989).

      Foreword by Ronald L. Graham and Persi Diaconis and "a new postscript and a new bibliography."

11. *Aha! Insight* (W.H. Freeman, 1978).

    - *Aha! A Two Volume Collection* (Mathematical Association of America, 2006).

      Combined with *Aha! Gotcha*, with a new Preface.

12. *Mathematical Circus* (Borzoi/Knopf, 1979).

    - *Mathematical Circus* (Vintage, 1981).
    - *Mathematical Circus* (Mathematical Association of America, 1992).

      Preface by Donald Knuth and "a postscript from the author and a new bibliography."

13. *Science-Fiction Puzzle Tales* (Potter, 1981).

    Puzzles from *Isaac Asimov's Science Fiction Magazine*; foreword by Isaac Asimov

    - *Mathematical Puzzle Tales* (Mathematical Association of America, 2000).

      With a new Postscript, briefly updating chapters 5, 6, 8, 13, 21, 23, and 30.

14. *Aha! Gotcha: Paradoxes to Puzzle and Delight* (W. H. Freeman, 1982).

    - *Aha! A Two Volume Collection* (Mathematical Association of America, 2006).

      Combined with *Aha! Insight*, with a new Preface.

15. *Wheels, Life, and Other Mathematical Amusements* (W. H. Freeman, 1983).

16. *Baffling Brainteasers* (Davis Publications, 1983).

    This small book reprints 13 puzzle-tales from *Isaac Asimov's Science Fiction Magazine* that have appeared in other books.

17. *Puzzles from Other Worlds* (Vintage, 1984).

    Further puzzles from *Isaac Asimov's Science Fiction Magazine*.

18. *The Magic Numbers of Dr. Matrix* (Prometheus, 1985).

    - *The Magic Numbers of Dr. Matrix* (Dorset, 1990).

19. *Knotted Doughnuts and Other Mathematical Entertainments* (W. H. Freeman, 1986).

20. *Riddles of the Sphinx: and Other Mathematical Puzzle Tales* (Mathematical Association of America, 1987). (New Mathematical Library, vol. 32)

    Further puzzles from *Isaac Asimov's Science Fiction Magazine.*

21. *Time Travel and Other Mathematical Bewilderments* (W. H. Freeman, 1988).

22. *Penrose Tiles and Trapdoor Ciphers* (W. H. Freeman, 1989).

    Subtitled on jacket " ... and the Return of Dr. Matrix."

    - *Penrose Tiles and Trapdoor Ciphers and the Return of Dr. Matrix* (Mathematical Association of America, 1997).

      With new postscript (which is missing illustrations in the first printing.)

23. *A Gathering for Gardner* (W. H. Freeman, 1989).

    This is just a boxed set of the three previous volumes, in softcover:
    Knotted Doughnuts and Other Mathematical Entertainments
    Time Travel and Other Mathematical Bewilderments
    Penrose Tiles and Trapdoor Ciphers and the Return of Dr. Matrix

24. *Fractal Music, Hypercards, and More: Mathematical Recreations from Scientific American Magazine* (W. H. Freeman, 1992).

25. *My Best Mathematical and Logic Puzzles* (Dover, 1994).

26. *The Universe in a Handkerchief: Lewis Carroll's Mathematical Recreations, Games, Puzzles, and Word Plays* (Copernicus, 1996).

27. *The Last Recreations: Hydras, Eggs, and Other Mathematical Mystifications* (Copernicus, 1997). (Paperback in 2001)

28. Silvanus P. Thompson and Martin Gardner, *Calculus Made Easy* (St. Martin's Press, 1998).

29. *The Colossal Book of Mathematics* (Norton, 2001).

30. *A Gardner's Workout* (A. K. Peters, 2001).

    Collects most of the mathematical articles after *Scientific American*, with new postscripts.

31. *Martin Gardner's Mathematical Games: The Entire Collection of His Scientific American Columns* (Mathematical Association of America, 2005). (With the American Mathematical Society it was sold as an ebook in 2020.)

    On a searchable CD-rom with a booklet by Albers and Renz.

32. *The Colossal Book of Short Puzzles and Problems* (Norton, 2006).

    Edited by Dana Richards, with a new Introduction and Afterword by Gardner.

33. *Aha! A Two Volume Collection* (Mathematical Association of America, 2006).

    Combined edition of *Aha! Insight* and *Aha! Gotcha*, with a new Preface.

1.2   Marketed Items

1. *Hexapawn: A Game You Play to Lose* (IBM, 1969).

   A 26.5" by 14.5" cardboard sheet, folded, with spinner.

2. *The Paradox Box* (W.H. Freeman, 1975).

   Filmstrips for the classroom; repurposed for *Aha! Gotcha*.

3. *Aha! Insight Box* (W.H. Freeman, 1978).

   Filmstrips for the classroom; repurposed for *Aha!*.

4. *The Game of Solomon* (Kadon Enterprises [Pasadena, MD], 1985).
5. *Lewis Carroll's Chess Wordgame* (Kadon Enterprises [Pasadena, MD], 1991).
6. *Visual Brainstorms 2* (Binary Arts [Alexandria, VA], 1997).

   Wrote a separate pamphlet, entitled "Introduction," to be included in this box of cards with puzzles on them. Many of the puzzles in this box, and in the first box as well, were taken from Gardner's books.

7. *Scientific American Mathematical Games* (Pomegranate Communications Inc [Dover], no date).
   - *Scientific American Mathematical Games: Volume 2* (Pomegranate Communications Inc [Dover], no date).
     Deck of cards with one puzzle on each using new illustrations.

8. *Classic Mind-Bending Puzzles* (Pomegranate Communications Inc [Dover], 2007).

   A puzzle-a-day calendar.

9. *Classic Mind-Bending Puzzles* (Pomegranate Communications Inc [Dover], 2008).

   A puzzle-a-day calendar.

1.3   Books Edited

1. Sam Loyd, *Mathematical Puzzles of Sam Loyd, Volume 1* (Dover, 1959).
2. Sam Loyd, *Mathematical Puzzles of Sam Loyd, Volume 2* (Dover, 1960).

   - Sam Loyd, *More Mathematical Puzzles of Sam Loyd,* (Dover, 1960). This later retitling does not have a different date.

3. H. E. Dudeney, *536 Puzzles and Curious Problems* (Scribners, 1967).

   - H. E. Dudeney, *536 Puzzles and Curious Problems and Puzzles* (Barnes and Nobles Books, 1995).
   - H. E. Dudeney, *536 Puzzles and Curious Problems* (Dover, 2016).

4. Boris Kordemsky, *Moscow Puzzles: 359 Mathematical Recreations* (Scribners, 1972).

   Using a translation by Albert Parry.

   - Boris Kordemsky, *Moscow Puzzles: 359 Mathematical Recreations* (Dover, 1992).

     "Slightly altered, slightly corrected republication."

5. Kobon Fujimura, *The Tokyo Puzzles* (Scribners, 1978).

   Using a translation by Fumie Adachi

1.4   Book Introductions

1. Karl Menninger, *Calculators Cunning: The Art of Quick Reasoning* (Basic, 1961) 5-6.
2. Samuel Randlett, *Best of Origami* (Dutton, 1963) 1-2.
3. Y. Yoshino, *The Japanese Abacus Explained* (Dover, 1963) v-xii.
4. Hugo Steinhaus, *One Hundred Problems in Elementary Mathematics* (Basic, 1964) [1-5].

   - Hugo Steinhaus, *One Hundred Problems in Elementary Mathematics* (Dover, 1979) [1-5].

5. Harold Jacobs, *Mathematics: A Human Endeavor* (Freeman, 1970) ix-x.
6. Pierre Berloquin, *100 Geometric Games* (Scribners, 1976). also (Dover, 2015)

   Same introduction used in several books.

   - Pierre Berloquin, *100 Numerical Games* (Scribners, 1976) vi-vii. also (Barnes & Noble, 1994) and (Dover, 2015).
   - Pierre Berloquin, *100 Games of Logic* (Scribners, 1976) ix-x. also (Barnes & Noble, 1995)
   - Pierre Berloquin, *100 Perceptual Puzzles* (Scribners, 1976) ix-x. also (Barnes & Noble, 1995)

7. Rick Brightfield, Glory Brightfield, *The Great Round the World Maze Trip* (Ballantine, 1977) 5-6.
8. Frank Harary, Editor, *Topics in Graph Theory* (New York Academy of Sciences, 1979) [5-6].
9. Mitsumasa Anno, *The Unique World of Mitsumasa Anno: Selected Works (1968-1977)* (Philomel Books, 1980) 4-5.

   - Mitsumasa Anno, *Anno's Unique World* (Collin's, 1980) 4-5.

10. Jim Moran, *Wonders of Magic Squares* (Vintage, 1981) xi-xii.
11. Raymond Smullyan, *Alice in Puzzleland* (Morrow, 1982) vii-x.

    - Raymond Smullyan, *Alice in Puzzleland* (Dover, 2011) viii-x.

12. Tom Werneck, *Die Zauberkugel* (Wilhelm Heyne Verlag, 1982) 6-8.

    In German and untranslated. The book is concerned with a single mechanical puzzle.

13. Harold Jacobs, *Mathematics: A Human Endeavor (Second Edition)* (W. H. Freeman, 1982) ix-xi. (Also Third Edition, 1994)

    Different introduction from the first edition, above.

14. Rudy Rucker, *The Fourth Dimension: Toward a Geometry of a High Reality* (Houghton Mifflin, 1984) ix-x.

    - Rudy Rucker, *The Fourth Dimension: A Guided tour of the Higher Universe* (Houghton Mifflin, 1985) ix-x.
    - Rudy Rucker, *The Fourth Dimension: Toward a Geometry of Higher Reality* (Dover, 2014) ix-x.

15. Stanislaw M. Ulam, *Science, Computers, and People: From the Tree of Mathematics* (Birkhauser, 1986) vii-x.

16. Jerry Slocum and Jack Botermans, *Puzzles Old & New: How to Make and Solve Them* (Seattle: University of Washington Press, 1986) 7-8.
17. E. T. Bell, *Mathematics: Queen and Servant of Science* (Mathematical Association of America, 1987) v-viii.
18. Robert Abbott, *Mad Mazes* (Bob Adams, 1990) 2.
19. Don Albers, et al. (Eds), *More Mathematical People* (Harcourt Brace Jovonavich, 1990) xi-xv.
20. Ian Stewart, *Another Fine Math You've Got Me Into* (W. H. Freeman, 1992) vii.
21. Steven Kahan, *At Last!! Encoded Totals Second Addition* (Baywood, 1994) xi-xii.
22. Alexander Soifer, *Colorado Mathematical Olympiad* (Center for Excellence in Mathematical Education, 1994) vii.
23. Terry Stickels, *Mind-Bending Puzzles, Volume 1* (Pomegranate, 1998) 3.
24. Beck, Anatole, Bleicher, Michael N., Crowe, Donald W., *Excursions into Mathematics: The Millennium Edition* (A. K. Peters, 2000) xix-xx.
25. Slocum, Jerry, *The Tangram Book* (Sterling, 2003) 7.

    Pictorial boards and dust jacket; a few copies have tangrams included.

26. Owen O'Shea, Underwood Dudley, *The Magic Numbers of the Professor* (Mathematical Association of America, 2007) vii-ix.
27. Jerry Slocum and Jack Botermans, *Het Ultieme Puzzleboek* (Terra, 2007) 9.

    • Jerry Slocum and Jack Botermans, *The World's Best Paper Puzzles* (Sterling, 2007) 7.

      This book is contained in a box with several punch-out physical puzzles. This represents only one chapter of the Dutch book, but was all they could get published.

28. Peter Moresca (Ed), *The Upside Down World of Gustave Verbeek* (Sunday Press Books, 2009) 7.
29. Persi Diaconis, Ron Graham, *Magical Mathematics* (Princeton University Press, 2012) ix-x.

1.5   Columns

1. "Puzzles - Tricks - Fun." *Uncle Ray's Magazine* (September 1946).

   Appeared monthly until March 1947. Illustrated by Gardner

2. "On the Light Side." *Science World* **1**, 1 (February 5, 1957).

   This appeared in volumes 1 through 5 ending on May 19, 1959. Each volume had 8 issues appearing every other week over one semester. Beginning with volume 2, Gardner used the pseudonym George Groth.

3. "Mathematical Games." *Scientific American* **196**, 1 (January 1957).

   This series appeared monthly until December 1980, vol. 243, no. 6. It appeared in the even-numbered issues of 1981, vols. 244 and 245. Additional columns appeared in August and September of 1983, vol. 249, nos. 2 and 3, and June 1986, vol. 254, no. 6. See separate chapter for much more detail.

4. "What's the Answer." *Coronet* (August 1968).

   Continued until July 1969. Selections from *Puzzles of Sam Loyd*.

5. "Science Fiction Puzzle Tales." *Isaac Asimov's Science Fiction Magazine* **1**, 1 (Spring 1977).

   Continued until vol. 10, no. 11, November 1986. See separate chapter for much more detail.

6. *Science Digest* [New Series] (March 1989 – April 1990).

   "Science fiction puzzles concocted by mathemagician Martin Gardner" that were patterned on those in *Isaac Asimov's Science Fiction Magazine*. The column was short-lived because the journal only lasted two years.
   "The Toroids of Dr. Kloneflake," March 1989, pp. 98-101
   "The Jock Who Wanted to be Fifty," May 1989, pp. 99-102
   "The Erasing of Philbert the Fudger," July 1989, pp. 99-102
   "Valley of the Apes," September 1989, pp. 99-102
   "How Crock and Witson Cracked the Code," November 1989, pp. 99-102
   "The Explosion of Blabbage's Oracle," January 1990, pp. 98-101
   "It's Off to Shuffle We Go," April/May 1990, pp. 83-86

7. "Physics Trick of the Month." *The Physics Teacher* (September 1990).

   Published until October 2002; see the chapter "Physics Tricks" in the Magic part.

8. "Gardner's Gatherings." *Math Horizons* (November 1994).

   Published irregularly until the February 2003. The individual columns are given as separate article citations below and appeared in these issues: September 1994, November 1994, April 1995, November 1995, February 1996, April 1996, November 1996, April 1997, September 1997, February 1998, September 1998, November 1999, February 2001, February 2003. Seven of these were collected in *Gardner's Workout*; the remainder were similar to *Scientific American* material.

1.6   Articles

Entries from the *Scientific American* and *Isaac Asimov's Science Fiction Magazine* columns, that have appeared in books, are listed separately in their respective chapters. Articles that emphasize magic tricks are listed separately in the Magic part, such as his articles in *Scripta Mathematica*.

1. "A Puzzling Collection." *Hobbies* (September 1934) 8.
2. "Flexagons." *Scientific American* **195** (December 1956) 162-166.
3. "If You Think They're Easy, Look Out." *Washington Post and Times Herald* (February 21, 1957) D2.

   This was a newspaper column, "The District Line," that ran five of the nine puzzles that had just appeared in "Mathematical Games," virtually unedited.

4. "Topology: A Strange New Mathematics." *Science World* **1**, 4 (March 19, 1957) 7-9.
5. "The Laws of Chance." *Science World* **2**, 5 (November 19, 1957) 14-15.

   - "The Laws of Chance." *Science World* **8**, 4 (March 15, 1961) 14-15.

6. "What's In a Curve." *Science World* **3**, 1 (February 6, 1958) 7-9.
7. Martin George [Martin Gardner], "Cram Course." *Gent* (April 1958) 21, 69.

8. "[Research Problem]." *Recreational Mathematics Magazine* 2 (April 1961) 40-41.

9. "Some Comments on Probability and Statistics." In *Probability and Statistics: An Introduction through Experiments*, edited by Edmund C. Berkeley (Science Materials Center, 1961) 110-112.

   Chapter 12 in this booklet that accompanied a box of materials, was actually a long letter, detailing examples, that Gardner wrote to the author.

10. Walter Stacey [Martin Gardner], "Ice Breakers." *Rogue* (January 1962) 35-37.

    A collection of simple tricks and stunts, some off-color droodles, in a men's magazine, a confirmed pseudonymous effort.

11. Martin Parrish [Martin Gardner], "Bar-oodles." *Rogue* (January 1962) 29, 80.

    Gardner independently invented "droodles" in the 1930's, so this set of six off-color droodles, in a men's magazine, found in his files, is *probably* a pseudonymous effort.

12. "Martin Gardner Presents a Pair of Puzzlers." In *The Fireside Calendar and Engagement Book* (Simon and Schuster, 1962). Unpaginated, opposite June 3 with answers at the end.

13. "Want to Be Fuddled." *Reader's Digest* **83** (October 1963) 37-38, 41.

14. "Bewitched, Bothered and Befuddled." *Reader's Digest* **85** (September 1964) 21-22, 44.

15. [Assortment of Mathematical Games] ([1964]).

    This packet was sent to subscribers of *Scientific American* as a "Season's Greetings". It contained an instruction book for 10 cards, each with a puzzle, games, illusion, or activity, plus a tetra-tetra-flexagon.

16. "A Dumbbell's Guide to the Space Age." *Esquire* (May 1965) 122-125.

    Contains several physics tricks. No by-line, and heavily edited by *Esquire*.

17. *The Platonic Solids* (1966).

    This lengthy article only appeared in a "Season's Greeting" card, a 15" by 20" folded sheet, sent to subscribers of *Scientific American*. "In the five sheets of heavy paper herewith you will find, in two dimensions, the five regular solids ready to be folded up ...On successful completion of this operation you will have five baubles to adorn a Christmas tree."

18. "[Various puzzles]." In *Teasers and Tests*, edited by Reader's Digest (Funk and Wagnalls, 1967) 37-38, 68, 115-116, 140, 154-155, 171.

Contains "Bewitched, Bothered or Befuddled?," "Want to Be Fuddled," and "Four Fun.". Also released as a promotional pamphlet with the same title, *Teasers and Tests* in 1967 containing only "Bewitched, Bothered or Befuddled?", p. 4.

19. "The Enigma That Is Pi." *Panorama (Chicago Daily New)* (January 7, 1967) 5.

Excerpted from *New Mathematical Diversions*

20. "It's More Probable Than You Think." *Denver Post: Empire Magazine* (October 8, 1967) 22.

- "It's More Probable Than You Think." *Reader's Digest* **191** (November 1967) 107-110.

21. "Rockefeller Called 'Best Balanced' of 1968 Presidential Front-Runners." *News* (November 8, 1967).

A press release "from the Inner Sanctum of Simon Schuster" written by Dr. Matrix/Gardner to advertise *The Numerology of Dr. Matrix*, 3 pages.

22. "A Magic Square for the NEW YEAR." *Mathematics Teacher* **61**, 1 (January 1968) 18.

23. "[Private Communication]." *Mathematics of Computation* **23** (1969) 548.

24. "Palindromes Revisited: The Front and Back of It." *Washington Post* (September 13, 1970) B2.

25. "More About Magic Star Polygons." *American Mathematical Monthly* **79** (1972) 396-397.

A comment on problem E 2265.

26. "[Paradoxes]." In *Vicious Circles and Infinity: An Anthology of Paradoxes*, edited by P. Hughes and G. Brecht (Doubleday, 1975) 39-42, 51-52, 57-59.

27. "[Acceptance Letter]." *American Mathematical Monthly* **83**, 10 (December 1976) 834.

For an honorary life membership in the Mathematical Association of America.

28. "My Ten Favorite Brainteasers." *Games* (January/February 1978) 16.

- *My Best Mathematical and Logic Puzzles* (Dover, 1994).

  A selection of these brainteasers appears amongst puzzles 55 to 66.

- *Games Magazine Big Book of Games* (Workman Publ Co, 1984) 130-131.

29. "How Logical Are You." *Reader's Digest* **112** (January 1978) 139-140.

30. *Oddities in Words, Pictures and Figures* (Reader's Digest, 1978).

This thin anthology contains unspecified contributions from the first two *Scientific American* books.

31. "10 Tricky Brainteasers." *Games* (November/December 1978) 18-19.

- *My Best Mathematical and Logic Puzzles* (Dover, 1994).

  A selection of these brainteasers appears among puzzles 55 to 66.

32. "Mathematical Circus." *Games* (September/December 1979) 15-16, 71.

33. "[Various puzzles]." In *Tests and Teasers*, edited by Reader's Digest (Berkley, 1980) 6-7, 18, 23, 41-43, 78, 120-121.

An updating of the 1967 "Teasers and Tests." Contains "Bewitched, Bothered or Befuddled?" and "How Logical are You?", in addition to some short puzzles from *Scientific American*. These excerpts were also released in 1980 in two different promotional pamphlets with the same title, *Tests and Teasers*, one with a red cover the other with a white cover.

34. "Det enkle og det overraskende [Danish]." In *Dobbeltmasken: Piet Hein 75 ar*, edited by Jarl Borgen (Borgen Forlag, 1980) 12-16.

35. "A Sample Gardner Delight." *Los Angeles Times* (December 12, 1981) A21.

36. "Tetrahexes." *Arithmetic Teacher* **29**, 6 (February 1982) 5.

A problem taken from *Mathematical Magic Show*

37. "Short Stuff." *Arithmetic Teacher* **29**, 8 (April 1982) 9, 49.

    Five problems taken from *Mathematical Circus*

38. "Puzzling Paradoxes." *Games* (September 1982) 18, 85.

    Tie-in with *The Paradox Box*

39. "My Favorite Funny Quickies." *Four-Star Puzzler* 28 (April 1983) 1,4.

    - "From the Flim-Flam File." In *The Book of Sense and Non-sense Puzzles*, edited by Ronnie Shushan (Workman Publ, 1985) 32-33, 148.

    This *Games Magazine* book contains an abridgment of the original.

40. "[A comment on 'The Fractal Geometry of Mandelbrot']." *College Mathematics Journal* **15** (1984) 117-118.

41. "A Connoisseur's Guide to Practical Jokes." *Games* (April 1984) 18-21.

42. "FLIP, the Psychic Robot." *Games* (October 1984) 22.

    - *GW* (2001) 21-23.
    - "FLIP, the Psychic Robot." *Games* (October 2010) 65, 58.

43. "Beauty, Wonder, and Fun." In *A Museum of Fun: Part II (Catalogue)*, edited by The Asahi Shimbun (Topran (Japan), 1984) 14-15.

    A Japanese translation appears on pp. 12-13.

    - "Beauty, Wonder, and Fun." *Games* (December 1984) 12-15.

44. "Aha [Spanish]." *Humor y Juegos* 48 (1984) 24-25.

45. "Dividiendo Pi en Millones." *Cacumen* [Spanish] **3**, 2 (March 1985) 12-14.

    - *W&W* (1989) 81-89.

46. "The Traveling Salesman's Travail." *Discover* (April 1985) 87-90.

    - *W&W* (1989) 90-101.

47. "Solomon." *Games* (April 1985) 46-47.

48. "The Soul of an Old Machine." *Discover* (May 1985) 36-43.

    - "The Abacus." In *W&W* (1989) 102-106.

49. "The Binary System." In *The Cyberdeck*, edited by Paul Swinford (Haines House of Cards, 1986) 1-8.

    This has the December 1960 *Scientific American* column used as chapter 1 of this book of magic tricks inspired by that column. Swinford added a "Gardner Addenda" about that column.

50. "Playing the Numbers." *Washington Post Education Review* (April 5, 1987) ER6.

    - *W&W* (1989) 151-152.

51. "Primes in Arithmetic Progression." In *1988 Mathematical Sciences Calendar*, edited by Nicholas Rose (Rome Press, 1987) 4, 6.

    - *GW* (2001) 175-179.

52. "Primes Magic Squares in General." In *1988 Mathematical Sciences Calendar*, edited by Nicholas Rose (Rome Press, 1987) 6, 8.

    - *GW* (2001) 181-185.

53. "Smith Numbers." *REC Newsletter: Recreational & Educational Computing* **3**, 1 (January 1988) 3-4.

54. "[Response]." *Notices of the American Mathematical Society* **34**, 6 (October 1987) 875-876.

    A short response to being awarded the Steele Prize for expository writing.

55. Martin Gardner and Frank Harary, "The Propositional Calculus with Directed Graphs." *Eureka* 48 (March 1988) 34-40.

    - *GW* (2001) 25-33.

56. "Crack the Problem; Win Fame." *New York Times* (May 25, 1988) A27.

57. "Four Curiosities for the New Year." In *1989 Mathematical Sciences Calendar*, edited by Nicholas Rose (Rome Press, 1988) 4.

58. Fan Chung, Martin Gardner, and Ronald L. Graham, "Steiner Trees on a Checkerboard." *Mathematics Magazine* **62**, 2 (April 1989) 83-96.

    - *GW* (2001) 39-59.

59. "Behind the Eight Ball." *Games* (October/November 1989) 18-19.

60. Martin Gardner and Lee Sallows, "Rectangling a Pi-Omino and Lord Zigzag's Tour." *REC Newsletter: Recreational & Educational Computing* **4**, 8 (November 1989) 6-7.

61. "Pi-Omino Dissection and Challenge." *REC Newsletter: Recreational & Educational Computing* **4**, 8 (November 1989) 8.
62. "The Opaque Cube Problem." *Cubism for Fun* 23 (March 1990) 15.

    • *GW* (2001) 3-4.

63. "Tiling the Bent Tromino with *N* Congruent Shapes." *Journal of Recreational Mathematics* **22**, 3 (1990) 185-191.

    • *GW* (2001) 61-71.

64. "Covering a Cube with Congruent Polygons." *Cubism for Fun* 25, Part 1 (December 1990) 13.

    • *GW* (2001) 73-75.

65. "The Opaque Cube Again." *Cubism for Fun* 25, Part 1 (December 1990) 14.

    • *GW* (2001) 4-7.

66. "Calculator Tricks to Amuse and Bemuse: Variations on the 12345679 Trick." *REC Newsletter: Recreational & Educational Computing* **6**, 3 (May 1991) 10.

    • *GW* (2001) 83-84.

67. "Sam Loyd." *Games* (July 1991) 16.
68. Lee Sallows, Martin Gardner, Richard K. Guy, and Donald Knuth, "Serial Isogons of 90 Degrees." *Mathematics Magazine* **64**, 5 (December 1991) 315-324.

    • *GW* (2001) 215-229.

69. "Two 3-Point Tiling Problems." *Cubism for Fun* 28 (April 1992) 32-35.

    • *GW* (2001) 143-147.

70. Martin Gardner and Andy Liu, "A Royal Problem." *Quantum* (July/August 1993) 30-31, 60.
71. "Cornering the King." *REC Newsletter: Recreational & Educational Computing* (July/August 1992) 5.

    • *GW* (2001) 109-115.

72. "Playing Physics." *Games* (October 1993) 18-19, 42.

    • "Playing Physics." *Games* (May 2012) 14-15, 58.

73. Peter Brugger and Martin Gardner, "Perseveration in Healthy Subjects: An Impressive Classroom Demonstration for Educational Purposes." *Perceptual and Motor Skills* **78** (1994) 777-778.

    A simple calculating stunt is recommended for classroom use.

74. "The Big Holdup." In *Mystery Maze*, edited by James Barry Et al [Eds] (Nelson Canada, 1994) 78-89.

    From *Aha*.

75. "Six Challenging Dissection Tasks." *Quantum* (May/June 1994) 28-29.

    • *GW* (2001) 121-128.

76. "Delicious Dissections." *Math Horizons* **2**, 1 (September 1994) 18-19.

77. "Toroidal Currency." *Quantum* (September/October 1994) 52-53.

    • *GW* (2001) 117-119.

78. "Puzzling with Martin Gardner." *Scientific American* **273**, 6 (December 1995) 26, 41.

79. Dan Witowski, *Funhouse Mirrors* (Abracadazzle, 1995).

    Big book with mirrored insets which contains unspecified "special material by Trip Johnson and Martin Gardner."

80. "Lewis Carroll's Sleepless Nights." *Quantum* (March/April 1995) 40-41.

    • *GW* (2001) 129-132.

81. "Around the Solar System." *Math Horizons* **2**, 4 (April 1995) 22-23.

82. "Dr. Matrix on the Wonders of 8." *Quantum* (July/August 1995) 43-44.

    Originally appeared as a press release, in April 1989, from the American Mathematical Society, as part of packet to commemorate the AMS centenary.

83. "The Game of Hip." *Math Horizons* **3**, 2 (November 1994) 20-21.

84. "The Recreational Mathematics of Piet Hien." In *Sin Egen: Piet Hien 90 Ar* (Borgen, 1995) 57-62.

85. "The Magic of 3×3." *Quantum* (January/February 1996) 24-26.

    • *GW* (2001) 157-165.

86. "The Ant on a $1 \times 1 \times 2$." *Math Horizons* **3**, 3 (February 1996) 8-9.

    • *GW* (2001) 139-141.

87. "The Latest Magic." *Quantum* (March/April 1996) 60.

    A follow-up to the previous article; also, with solutions for that article.

88. "Talkative Eve." *Math Horizons* **3**, 4 (April 1996) 18-19.

- "Talkative Eve." *Math Horizons* **18**, 1 (September 2010) 6-7.

89. "Numbers of the Beast." *Games* (April 1996) 48.

One of five "Beguilers"

90. "Ridiculous Questions." *Math Horizons* **4**, 2 (November 1996) 24-25.

91. "Lucky Numbers and 2187." *Mathematical Intelligencer* **19**, 2 (1997) 26-29.

- *GW* (2001) 149-156.

92. "The Square Root of Two = 1.41421 35623 73095 ...." *Math Horizons* **4**, 4 (April 1997) 5-8.

- *GW* (2001) 9-19.

93. "Some Surprising Theorems about Rectangles in Triangles." *Math Horizons* **5**, 1 (September 1997) 18-22.

- *GW* (2001) 203-213.
- "Some Surprising Theorems about Rectangles in Triangles." In *The Edge of the Universe*, edited by Deanna Haunsberger and Stephen Kennedy (Mathematical Association of America, 2006) 66-69.

94. "My Favorite New Brain Teasers." *Games* (October 1997) 46-47.

95. "An Unusual Magic Square and a Prize Offer." *Cubism for Fun* 45 (February 1998) 8.

96. "Some New Discoveries about 3×3 Magic Squares." *Math Horizons* **5**, 3 (February 1998) 11-13.

- *GW* (2001) 167-173.
- "Some New Discoveries about 3×3 Magic Squares." In *The Edge of the Universe*, edited by Deanna Haunsberger and Stephen Kennedy (Mathematical Association of America, 2006) 74-76.

97. "A Quarter-Century of Recreational Mathematics." *Scientific American* **279**, 2 (August 1998) 68-75.

- "Les Jeux Mathématiques." *Pour la Science* **279**, 252 (Octobre 1998) 86-94.
- *GW* (2001) 191-201. The original was "a heavily revised and cut version."

98. "The Unattacked Cells Problem." *Journal of Recreational Mathematics* **29**, 2 (1998) 155-157. Problem 2373 followed by the "Solution by the proposer"

99. "The Asymmetric Propeller." *The College Mathematics Journal* **30**, 1 (January 1999) 18-22.

   - *GW* (2001) 249-255.
   - "The Asymmetric Propeller." In *Martin Gardner in the Twenty-First Century*, edited by Michael Henle and Brian Hopkins (Mathematical Association of America, 2012) 3-6.

   With follow-up articles by others.

100. "Dominono." *Games* (April 1999) 53, 44.

   - "Dominono." *Computers and Mathematics with Applications* **39** (2000) 55-56, 171.
   - *GW* (2001) 187-189.

101. "Tic-Tac-Toe Played as a Word Game." *Word Ways* **32**, 3 (August 1999) 205.

   A query answered by D. E. Knuth November 1999, p. 262, with an acknowledgment by Gardner.

   - "Tic-Tac-Toe Played as a Word Game." *Word Ways* **43**, 3 (August 2010) 227.

102. "Chess Queens and Maximum Unattacked Cells." *Math Horizons* **7**, 2 (November 1999) 12-16.

   - *GW* (2001) 257-267.

103. "The Universe in a Handkerchief." *MD Computing* (March/April 2000) 80.

   Excerpted from *Universe in a Handkerchief*.

104. "Lavinia Seeks a Room and Other Problems." *MD Computing* (May/June 2000) 72.

   Excerpted from *Last Recreations*.

105. "Modeling Mathematics with Playing Cards." *The College Mathematics Journal* **31**, 3 (May 2000) 173-177. Answers in next issue, p. 280

   - *GW* (2001) 241-248.
   - "Modeling Mathematics with Playing Cards." In *Martin Gardner in the Twenty-First Century*, edited by Michael Henle and Brian Hopkins (Mathematical Association of America, 2012) 221-266.

106. "Some New Results on Magic Hexagrams." *The College Mathematics Journal* **31**, 4 (May 2000) 274-280.

- "New Results on Magic Hexagrams." In *UTB* (2003) 74-82.
- "Some New Results on Magic Hexagrams." In *Martin Gardner in the Twenty-First Century*, edited by Michael Henle and Brian Hopkins (Mathematical Association of America, 2012) 159-166.

With follow-up results by others.

107. "From Oz to Earth." *Math Horizons* **8**, 1 (September 2000) 18-19.

Excerpt from *Visitors from Oz* with a sidebar from *Scientific American*.

108. "The Magic Mirror." *Games* (September 2000) 4.

The puzzle itself appears on the cover.

109. "Pieces at Peace." *Games* (September 2000) 58, 47.
110. "Fit for a Queen." *Games* (October 2000) 62, 47.

- "Fit for a Queen." *Games* (February 2013) 79, 63.

111. "Math with Scissors." *Scientific American Explorations* (Fall 2000) 10, 12, 54.

- "Matemáticas con Tijeras." *Scientific American* [Latinoamerica]*Exploraciones* (Abril 2003) 6-7.
- "Fun with Möbius Bands." In *UTB* (2003) 57-60.
- "Fun with Möbius Bands." *The Ojai Orange* 36 (March 2005). ("John Wilcock's personal magazine")(Part of "The Modest Genius of Martin Gardner".)

112. "Is Mathematics 'Out There'?" *Mathematical Intelligencer* **23**, 1 (2001) 7-8.

- *UTB* (2003) 61-63.

113. "Some New Results on Nonattacking Chess Tasks." *Math Horizons* **8**, 3 (February 2001) 10-12.

- "Some New Results on Nonattacking Chess Tasks." In *The Edge of the Universe*, edited by Deanna Haunsberger and Stephen Kennedy (Mathematical Association of America, 2006) 178-180.

114. "Calendar Conundrums." *Games* (September 2001) 87, 63.
115. "Rear View." *Games* (October 2001) 79, 63.
116. "A Surprising Match." *Games* (November 2001) 78, 63.

117. "Mel Stover." In *Puzzlers' Tribute: A Feast for the Mind*, edited by David Wolfe and Tom Rodgers (A. K. Peters, 2002) 29-32.

118. Jeremiah Farrell, Martin Gardner, and Thomas Rodgers, "Configuration Games." In *Tribute to a Mathemagician*, edited by Barry Cipra, Erik Demaine, Martin Demaine, Tom Rodgers (A K Peters, 2005) 93-99.

119. Mamikon Mnatsakanian, Gwen Roberts, and Martin Gardner, "Networds." *Games* (May 2005) 65-68.

- Mamikon Mnatsakanian, Gwen Roberts, and Martin Gardner, "Networds." In *A Lifetime of Puzzles*, edited by Erik D. Demaine, Martin L. Demaine, and Tom Rodgers (A. K. Peters, 2008) 265-272.

120. "Transcendentals and Early Birds." *Math Horizons* **13**, 2 (November 2008) 5-6.

- *JFH* (2008) 89-92.
- "Transcendentals and Early Birds." In *Martin Gardner in the Twenty-First Century*, edited by Michael Henle and Brian Hopkins (Mathematical Association of America, 2012) 37-38.

121. "Dr. Matrix on Little Known Fibonacci Curiosities." *Journal of Recreational Mathematics* **34**, 3 (2005-2006 [copyright 2008]) 183-190.

- *T&F* (2009) 106-113.

122. "L-Tromino Tiling of Mutilated Chessboards." *College Mathematics Journal* **40**, 3 (2009) 162-168.

- *T&F* (2009) 114-123.
- "L-Tromino Tiling of Mutilated Chessboards." In *Martin Gardner in the Twenty-First Century*, edited by Michael Henle and Brian Hopkins (Mathematical Association of America, 2012) 127-134.

123. "Puzzle Time!" *Wired* **17**, 5 (May 2009) 40 (answer on-line).

A toothpick puzzle solicited for the "The Mystery Issue."

124. "Martin Gardner: A Look Back." *Games World of Puzzles* (February 2015) 40-41, 76.

1.7 Book Reviews

1. "Numbers Always Count." *New York Times Book Review* (February 4, 1962) 24.

   The Last Problem, E.T. Bell

   - *O&S* (1983) 251-253.

2. *New York Herald Tribune* (August 19, 1962) 8.

   The Mathematical Magpie, Clifton Fadiman
   Mathematics for Pleasure, Oswald Jacoby and W.H. Benson

   - *O&S* (1983) 254.

3. "Paperbacks in Review: Mathematics." *New York Times Book Review* (November 25, 1962) 34.

   Short reviews of 25 paperbacks of expository mathematics.

   - "Publishers Add Edition on Math." *New York Times* [Western Edition] (February 20, 1963) K8.

     Abridged version.

4. "Publishers Add Editions on Math." *New York Times Book Review* (February 20, 1963) 8.

   Short reviews of 12 paperbacks of expository mathematics.

5. "From Euclid's Axioms to Gödels Proof." *Library of Science* (1963). Advertising brochure.

   Mathematical Recreations and Essays, W. W. Rouse Ball
   Gentle Art of Mathematics, Daniel Pedoe

6. "Topology." *Library of Science* (ca. 1964). Advertising brochure.

   Intuitive Concepts in Elementary Topology, B.H. Arnold
   Experiments in Topology, Stephen Barr

7. "Mathematical Puzzles and Paradoxes." *Library of Science* (1965). Advertising brochure.

   Puzzles and Paradoxes, Thomas H. O'Beirne
   A Miscellany of Puzzles, Stephen Barr

8. "Polyominoes and Mathematical Magic." *Library of Science* (1965). Advertising brochure.

   Polyominoes, Solomon W. Golomb
   Mathematical Magic, William Simon

9. *Library of Science* (1965). Advertising brochure.

   Famous Problems in Mathematics, H. Tietze

10. *Library of Science* (1969). Advertising brochure.

    Excursions in Geometry, C.S. Ogilvy

11. "A Revised Edition of a Modern Classic." *Library of Science* (1969). Advertising brochure.

    Introduction to Geometry, H.S.M. Coxeter

12. "The Joys of Mathematics." *Library of Science* (1969). A sidebar of recreational mathematics included.

    Mathematical Snapshots, H. Steinhaus
    Second Miscellany of Puzzles, Stephen Barr

13. "How Man Learned That 2+2=4 and More." *Washington Post Book World* (November 16, 1969) 5. For several years the *Chicago Tribune* and the *Washington Post* had the same book review pull-out section. For simplicity only the *Post* will be cited throughout.

    Number Words and Number Symbols, Karl Menninger

14. *Library of Science* (ca. 1970). Advertising brochure.

    Essays in Cellular Automata, A.W. Burks

15. *Library of Science* (1972). Advertising brochure.

    Tomorrow's Math, C. Stanley Ogilvy

16. *Library of Science* (1974). Advertising brochure.

    Symmetry in Art and Science, A.V. Shubnikov and V.A. Kopstik

17. "The Labyrinthian Way." *New York Times Book Review* (July 27, 1975) 10.

    A review of 28 maze books.

18. "Is Mathematics for Real?" *New York Review of Books* (August 13, 1981) 37-40.

The Mathematical Experience, Philip J. Davis and Reuben Hersch

- *O&S* (1983) 339-351.
- *TNIL* (1996) 280-293.

19. "On Another Book." *Isaac Asimov's Science Fiction Magazine* (February 1982) 21-22.

Infinity and the Mind, Rudy Rucker

- *WW&FL* (1996) 167-168.

20. "Some Really Hot Numbers." *Washington Post Book World* (March 4, 1984) 7.

Bridges to Infinity, Michael Guillen

- *W&W* (1989) 172-174.

21. "Quiz Kids." *New York Review of Books* (March 15, 1984) 23-25.

The Great Mental Calculators, Steven Smith

- *W&W* (1989) 179-185.

22. "The Fun of 'If ... Then'." *New York Times Book Review* (November 10, 1985) 52.

Anno's Hat Tricks, Akihiro Nozaki

- *W&W* (1989) 215-217.

23. "Count Up." *New York Review of Books* (December 3, 1987) 34-36.

To Infinity and Beyond, Eli Maor
Mind Tools, Rudy Rucker

- *W&W* (1989) 247-255.
- *TNIL* (1996) 50-58.

24. "Making Masters into Pawns of Life." *News and Observer* [Raleigh, NC] (December 4, 1988) 4D.

Searching for Bobby Fischer, Fred Waitzkin

- *WJ* (2000) 146-148.

25. "Beauty in Numbers." *New York Review of Books* (March 16, 1989) 26-28.

    Mathematics: The New Golden Age, Keith Devlin
    Labyrinths of Reason, William Poundstone

    - *WW&FL* (1996) 186-192.

26. "Theorems Floated in His Head." *News and Observer* [Raleigh, NC] (June 2, 1991).

    The Man Who Knew Infinity, Robert Kanigel

    - *WW&FL* (1996) 222-224.

27. *American Scientist* **80** (March/April 1992) 199-200.

    The Mathematics of Great Amateurs, Second Edition, Julian Lowell Coolidge

28. "Making No Apologies." *Nature* **358** (July 2, 1992) 28-29.

    The Art of Mathematics, Jerry King

    - *WW&FL* (1996) 240-241.

29. "A-Symmetry." *New York Review of Books* (December 3, 1992) 33-35.

    Fearful Symmetry: Is God a Geometer?, Ian Stewart and Martin Golubitsky
    Symmetry in Chaos, Michael Field and Martin Golubitsky
    M. C. Escher: Visions of Symmetry, Doris Schattschneider
    Wordplay: Ambigrams and Reflections on the Art of Ambigrams, John Langdon

    - *TNIL* (1996) 4-12.

30. "Can You Solve This?" *Washington Post Book World* (May 8, 1994) 19.

    A review of six puzzle books for children and adults.

31. "Mental Tunnels." *Nature* **374** (March 2, 1995) 25.

    Inevitable Illusions: How Mistakes of Reason Rule Our Minds, M. Piattelli-Palmarini

    - *WW&FL* (1996) 258-260.

32. *American Scientist* **83** (July/August 1995) 381.

   The Lighter Side of Mathematics, R. K. Guy and R. E. Woodrow (Eds)

33. *American Scientist* **84** (March/April 1996) 191-192.

   Lion Hunting and Other Mathematical Pursuits, G. L. Alexanderson and D. H. Mugler

   - *GW* (2001) 271-274.

34. *Los Angeles Times Book Review* (May 18, 1997) 5.

   Achilles in the Quantum Universe, Richard Morris
   The Invention of Infinity, J. V. Field

   - *GW* (2001) 275-279.

35. "Mathematical Realism and Its Discontents." *Los Angeles Times Book Review* (October 12, 1997) 8-9.

   What is Mathematics, Really?, Reuben Hersh
   The Number Sense, Stanislas Dehaene

   - *WJ* (2000) 217-227, 228-234.
     The review was split into two chapters, one for each book.

36. "The New New Math." *The New York Review of Books* (September 24, 1998) 9-12.

   Multicultural and Gender Equity in the Mathematics Classroom, Janet Trentacosta
   Focus on Algebra, Randall I. Charles, et al.
   Life by the Numbers, narrated by Danny Glover

   - *GW* (2001) 301-319.

37. "Magic Squares Cornered." *Nature* **395** (September 17, 1998) 216-217.

   Most-Perfect Pandiagonal Magic Squares: Their Construction and Enumeration, Kathleen Ollerenshaw and David Bree

   - *GW* (2001) 285-290.
   - *Wordways* **52**, 3 (2019). article 5, within the article "Cubic Magic," by Jeremiah Farrell

38. "It All Adds Up." *Los Angeles Times Book Review* (November 8, 1998) 7.

    The Number Devil, Hans Magnus Enzenberger

    - *GW* (2001) 291-294. The original version was a cut.

39. "Larger than Proof." *New Criterion* (December 2000) 77-80.

    Gödel: A Life of Logic, John L. Casti and Werner DePauli

    - "Kurt Gödel's Amazing Discovery." In *UTB* (2003) 67-73.

40. "Spheres of Influence." *Washington Post Book World* (June 24, 2001) 15.

    Flatterland, Ian Stewart

    - "Ian Stewart's *Flatterland.*" In *UTB* (2003) 64-66.

41. *Journal of Recreational Mathematics* **33**, 3 (2004/2005) 214. [Despite the official date, the copyright date is 2006. This a one paragraph summary of the longer 2007 review in *The College Mathematics Journal.*]

    PopCo, Scarlett Thomas

42. "Hotel Infinity." *New Criterion* (October 2005) 71-74.

    The Pea and the Sun, Leonard M. Wapner

    - *JFH* (2008) 83-88.

43. *The College Mathematics Journal* **38**, 3 (May, 2007) 241-242.

    PopCo, Scarlett Thomas

    - *JFH* (2008) 129-131.
    - "Review of *PopCo* by Scarlett Thomas." In *Martin Gardner in the Twenty-First Century*, edited by Michael Henle and Brian Hopkins (Mathematical Association of America, 2012) 287-289.

44. "Still Four." *New Criterion* **28**, 4 (December, 2009) 68-70.

    Mathematicians, Marianna Cook

45. "Abstract Adventuring." *New Criterion* **29**, 10 (June, 2010) 68-70.

    Mathematicians, Marianna Cook

1.8 Published Correspondence

1. *The Cryptogram* 2 (April 1932) 7.

   Suggestions for the fledgling American Cryptogram Association's magazine.

2. *Chicago Daily Tribune* (April 30, 1941) 15.

   Appeared in the column "Front Views and Profiles" by June Provines; "You are an elevator operator" puzzle.

3. "E 1309 [Dissection of a Regular Pentagram into a Square]." *American Mathematical Monthly* **65**, 3 (March 1958) 205.

   A solution appeared in the November 1959 issue, pp. 710-711; even though in was solved by "the proposer" an improved solution by H. Lindgren was given.

4. *The Mathematics Teacher* **53**, 5 (May 1960) 375.

   Traces Hawley's triangle dissection to Dudeney

5. *Recreational Mathematics Magazine* 8 (April 1962) 45.

   Gives attribution of the triangle-square dissection to Dudeney.

6. *Recreational Mathematics Magazine* **60**, 6 (October 1967) 587.

   Doodles of math words inspired by *MAD* magazine!

7. *Manifold* 8 (Autumn 1970) 51.

   Observes the rhombic dodecahedron is not Hamiltonian; mentions repunits.

8. *Manifold* 13 (Winter 1972/1973) 36.

   Gives a non-transitive dice example, due to Walter Penney

9. "Trade Secrets." *New York Review of Books* (November 8, 1984).

   Rejoinder about mental calculators.

   - *W&W* (1989) 186-187.

10. "Optimizing the Differences." *Journal of Recreational Mathematics* **17**, 3 (1984-1985) 215.

    Poses a card placement problem in the "Problems and Conjectures" section.

11. "Whoops." *Discover* (June 1985) 94.

    Corrections to the traveling salesman article.

12. "Update II." *Mathematics Teacher* **78** (September 1985) 409.

    Gives a proof for polygons using sliding matches.

13. "China, Gauss, and Pi." *The Sciences* [New York Academy of Sciences] (November/December 1985) 16.

    Gauss was possibly confused with Hobbes, who was naive of π.

14. "Problem: 341." *College Mathematics Journal* **18**, 1 (January 1987) 69.

    - "Jumping Pegs." *College Mathematics Journal* **20**, 1 (January 1989) 78-79. Solution to problem 341.

15. "The Pattern Updated." *Mathematics Teacher* **80** (January 1987) 5.

    Updates an article about Armstrong numbers.

16. "Chomp." *Mathematics Teacher* **80** (March 1987) 172.

    Adds to an article on the game Chomp [also see a letter by Richard K. Guy]

17. "An Infinity of Points." *New York Review of Books* (March 3, 1988) 45.

    Rejoinder about Cantor and infinity.

18. "The History of the Magic Hexagon." *Mathematical Gazette* (June 1988) 133.

19. "Squaring the Square." *The Sciences* [New York Academy of Sciences] (January/February 1990) 14,16.

    Corrects an assertion about Olivastro about geometric dissections

20. "Ask Marilyn." *Parade Magazine* (September 8, 1991) 7.

    Poses a conditional probability question.

    - Marilyn Vos Savant, *The Power of Logical Thinking* (St. Martin's Press, 1996) 24-25.

21. *Geombinatorics* **2** (1992) 17.

    Refers to material in *Logic Machines and Diagrams*.

22. George Groth [Martin Gardner], "Ask Marilyn." *Parade Magazine* (January 31, 1993) 10.

   Poses a conditional probability question.

   • Marilyn Vos Savant, *The Power of Logical Thinking* (St. Martin's Press, 1996) 55-56.

23. "Ask Marilyn." *Parade Magazine* (April 18, 1993) 12.

   A five-sided duel problem presented.

24. George Groth [Martin Gardner], "Ask Marilyn." *Parade Magazine* (November 28, 1993) 19.

   A parity problem is presented.

25. "Barr's Bar." *Skeptical Briefs* **4**, 1 (March 1994) 13.

   Brief note about Stephen Barr's length encoding idea.

26. "A Martin Gardner Gem is Unearthed." *REC Newsletter: Recreational and Educational Computing* **9** (October/November 1994) 6.

   Contributes a magic square puzzle found in a magic periodical of the 1930s.

27. 13118209147118414518 [Martin Gardner], "Ask Marilyn." *Parade Magazine* (March 12, 1995) 35.

   A coded message is presented with Gardner's name and address in code as well. Answer appeared March 19, p. 24.

28. 'Anonymous' [Martin Gardner], "Ask Marilyn." *Parade Magazine* (April 2, 1995) 16.

   An obfuscated description of a thimble.

29. "Distorted Images." *Scientific American* **273**, 5 (Nov. 1995) 8.

   Remarks about optical illusions.

30. "O. J. Numerology." *Word Ways* **28** (November 1995) 201.

   I. J. Matrix remarks on O. J. Simpson and the number 32.

31. "Geodesics and Magic Squares." *REC Newsletter: Recreational & Educational Computing* (February/March 1996) 8.

32. "Gardner: Try this Trick." *REC Newsletter: Recreational & Educational Computing* **11**, 1 (July/Agust 1996) 6.

    Explains the Steinmeyer nine-card trick

33. George Groth [Martin Gardner], "Ask Marilyn." *Parade Magazine* (November 3, 1996) 8.

    A logic problem is presented.

34. "The Unattacked Cells Problem." *Journal of Recreational Math* **28** (1996) 298.

    Problem 2373 in the "Problems and Conjectures" column.

35. "Leaves from the Deanery [Letters]." *Knight Letter* 57 (1998) 10.

    Discussion of a Dudeney problem in this publication of the Lewis Carroll Society of North America.

36. "Match This." *Mathematics Teacher* **91**, 4 (April,1998) 363-364.

    Using sliding matches to measure a polygon's angles

37. "Ask Marilyn." *Parade Magazine* (November 8, 1998) 20.

    A operations research toasting problem is presented.

38. "The New New Math." *New York Review of Books* (December 3, 1998) 61.

    A response to Timothy Craine

    - *GW* (2001) 318.

39. "Sandra's Math Block." *New York Review of Books* (December 3, 1998) 61.

    A response to Ruth Heaton

    - *GW* (2001) 319.

40. Armand T. Ringer [Martin Gardner], "Ask Marilyn." *Parade Magazine* (February 7, 1999) 100.

    The four bugs problem

41. "Ask Marilyn." *Parade Magazine* (May 23, 1999) 10.

    Place digits in a diagram

42. "Ask Marilyn." *Parade Magazine* (November 21, 1999) 14.

    Area of square in a circle in a square; answer November 28.

43. "Ask Marilyn." *Parade Magazine* (April 23, 2000) 18.

    Isosceles triangle inscribed in a square; answer April 30.

44. "Leaves from the Deanery [Letters]." *Knight Letter* 64 (Fall 2000) 12.

    Remarks about truth tables.

45. "Quickies: Q907." *Mathematics Magazine* **74**, 1 (February 2001) 67. Answer, A907, appears on page 71 of the same issue.

    A word-forming game isomorphic to tic-tac-toe.

46. "Ask Marilyn." *Parade Magazine* (March 13, 2005) 17, 18.

    Poses an original toothpick puzzle.

47. "Ask Marilyn." *Parade Magazine* (October 30, 2005) 16, 18.

    Distance problem with apparently too little data.

48. Armand T. Ringer [Martin Gardner], "Ask Marilyn." *Parade Magazine* (December 25, 2005) 16.

    Speed problem with apparently too little data.

49. "[Facsimiles of letters to Ross Eckler]." *Words Ways* **48**, 1 (February, 2015) 75-76, 79.

50. "[Facsimile of letter to Tapan Mukherjee]." *Words Ways* **53**, 4 (2020).

ODDITIES and CURIOSITIES
of WORDS and LITERATURE

(Gleanings for the Curious)

BY

C. C. BOMBAUGH

EDITED AND ANNOTATED

BY

MARTIN GARDNER

DOVER PUBLICATIONS, INC.
NEW YORK

Chapter 2

Recreational Linguistics

Wordplay is often found alongside recreational mathematics. It is certainly scattered throughout the *Scientific American* column. However, to avoid possible confusion, these citations have been separated out, to make it easier to understand their nature.

As a result of his work on Bombaugh's book he worked with leaders in the field like Dmitri Borgmann and Howard Bergerson. Using his relationship to Greenwood Press (*Journal of Recreational Mathematics*) he was the force behind the creation of *Word Ways: The Journal of Recreational Linguistics* in 1968, which lasted for over 50 years.

2.1   Books

1. *Colossal Book of Wordplay* (Puzzle Wright Press / Sterling, 2010).

   Published posthumously, with Ken Jennings polishing up the manuscript.

2.2   Books Edited

1. C. C. Bombaugh, *Oddities and Curiosities of Words and Literature* (Dover, 1961).
2. H. E. Dudeney, *300 Best Word Puzzles* (Scribners, 1968).

2.3 Book Introductions

1. Margaret Farrar, *The Crossword Puzzle Book (Series 107): 50th Anniversary Issue* (Simon and Schuster, 1974) [iii-iv].
2. John Langdon, *Wordplay: Ambigrams and Reflections on the Art of Ambigrams* (Harcourt Brace Javonovitch, 1992) xiii-xviii.

   The second edition (2005) has an introduction by Dan Brown.

3. Ross Eckler, *Making the Alphabet Dance* (St. Martin's Press, 1996) xix-xx.

2.4 Articles

1. "The Long and Short of It." *Children's Digest* (March 1952) 22-23.

   Two limericks with long and short last lines

2. "Kickshaws." *Word Ways* **14**, 1 (February 1981) 44-50.
   - *O&S* (1983) 188-194.
   - "Kickshaws." *Word Ways* **43**, 3 (August 2010) 191-197.

3. "Kickshaws." *Word Ways* **16**, 4 (November 1983) 238-245.
   - *W&W* (1989) 68-76.
   - "Kickshaws." *Word Ways* **43**, 3 (August 2010) 198-205.

4. "For Better or Verse." *Games* (January 1984) 24.

   "Seven Puzzle Poems"
   - *W&W* (1989) 77-80.

5. "Mathematics and Wordplay." *Word Ways* **26**, 1 (February 1993) 3-4.
   - *GW* (2001) 35-38.
   - "Mathematics and Wordplay." *Word Ways* **43**, 3 (August 2010) 212-213.

6. "Word Ladders – Lewis Carroll's Doublets." *Math Horizons* **2**, 2 (November 1994) 18-19.
   - "Word Ladders – Lewis Carroll's Doublets." *Mathematical Gazette* 487 (March 1996) 195-198, 203.
   - *GW* (2001) 133-137.

- "Word Ladders: Lewis Carroll's Doublets." In *The Edge of the Universe*, edited by Deanna Haunsberger and Stephen Kennedy (Mathematical Association of America, 2006) 22-23, 29.

7. "Linguistic Catches." *Word Ways* **40**, 1 (February 2007) 22-23.
8. "Word Play." *Word Ways* **40**, 4 (November 2007) 271-278..

2.5 Published Correspondence

1. "'Cause the Bible Told Me So." *Word Ways* **27** (May 1994) 88.

   A reprint of an exchange of letters that originally appeared in the *Time-News* [NC], in early 1994, that discuss Bible jokes.

   - "'Cause the Bible Told Me So." *Word Ways* **43** (August 2010) 210.

2. "The Poker Alphabet." *Word Ways* **27** (August 1994) 173-174.

   Contributes an 1899 alphabet poem

3. "[Letter]." *Word Ways* **48**, 1 (February 2015) 75-76.

   A facsimile of a letter written to Ross Eckler March 22, 1998

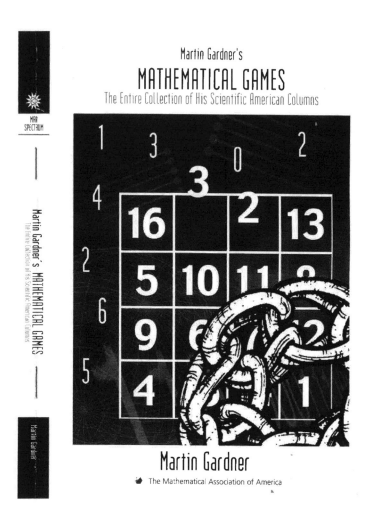

Figure 2.1: Cover to the CD collection of all 15 books

Chapter 3

The *Scientific American* Column

3.1 Introduction

Martin Gardner wrote nearly 300 columns for *Scientific American* over a period of more than 25 years. These columns have influenced countless individuals, not only mathematicians and puzzlers, but also people who found the essays a painless and inviting introduction to mathematics. The essays were characterized by their variety and the eclectic sources that Gardner drew upon. Another aspect of the column's appeal was the truly astonishing number of individuals—authors, researchers, and correspondents—that contributed material to the column.

Every column has been collected in a series of fifteen books. (Actually, one column, from October 1975, about ESP was collected in a more appropriate volume.) Most columns appeared in book form with addenda, citing new results and updates from correspondents. Many of the books have been revised and the addenda have been further updated. Further, some addenda grew so large that they were presented as separate chapters in the books; so a single column may correspond to one, two or even three chapters. These new chapters are designated NOT *SA*. Luckily, it was extremely rare for material to be removed from a column when collected into a book; mistakes were retained and corrected in the addendum. The titles of the original columns are not given, since they were written by the editors (not by Gardner) and served as descriptions, rather than titles.

It is tempting to conclude there were exactly 300 columns, however there were fewer, since during the final year only six columns appeared (alternating with Douglas Hofstadter's replacement column, "Metamagical Themas"). After these 294 columns, three more appeared sporadically under the "Mathematical Games" title. (Later essays, that could have similarly appeared in *Scientific American* but appeared elsewhere, were collected in *A Gardner's Workout,* A. K. Peters, 2001.) The December 1956 article on flexagons, that initiated the column, should be added to the count, as should the article that summarized the column ("A Quarter-Century of Recreational Mathematics," *Scientific American* , August 1998, pp. 68-75), giving a total of 299. However two other Gardner articles appeared in *Scientific American* that could easily have served as columns, and could be included in the count by the interested reader ("Logic Machines," March 1952, pp. 68-73 and "Can Time Go Backward?" January 1967, pp. 98-108).

Two very detailed indexes, a topics index and a name index, were constructed by hand and appear as appendices. These are new indexes but not the first attempts; previous indexes of the *Scientific American* columns are listed at the end of the "About Martin Gardner" appendix. The designators used here (such as FIRST) are the same as used in the appendixes.

3.2   Book Chapter Index

- [FIRST], *The Scientific American Book of Mathematical Puzzles and Diversions* (Simon and Schuster, 1959).

  - *Hexaflexagons & Other Mathematical Diversions: The First Scientific American Book of Puzzles & Games* (University of Chicago Press, 1988).

    "With a new afterword by the author, and an extensive new bibliography."

  - *Hexaflexagons, Probability, Paradoxes, and the Tower of Hanoi* (Cambridge University Press / Mathematical Association of America, 2008).

    With a new preface by the author, and "much fresh material."

- [SECOND], *The 2nd Scientific American Book of Mathematical Puzzles and Diversions* (Simon and Schuster, 1961).

  - *The 2nd Scientific American Book of Mathematical Puzzles and Diversions* (University of Chicago Press, 1987).

    "With a new postscript by the author."

  - *Origami, Eleusis, and the Soma Cube* (Cambridge University Press / Mathematical Association of America, 2008).

    Updated.

- [NEW], *New Mathematical Diversions from Scientific American* (Simon and Schuster, 1966).

  - *New Mathematical Diversions from Scientific American* (University of Chicago Press, 1983).

    "Phoenix edition."

  - *New Mathematical Diversions from Scientific American* (Mathematical Association of America, 1995).

    "With a postscript and a new bibliography."

  - *Sphere Packing, Lewis Carroll and Reversi* (Mathematical Association of America / Cambridge University Press, 2009).

    Updated

- [DRMATRIX], *The Numerology of Dr. Matrix* (Simon and Schuster, 1967).

  - *The Incredible Dr. Matrix* (Scribners, 1976).
    Expands the previous book.

  - *The Magic Numbers of Dr. Matrix* (Prometheus, 1985).
    Expands both books with new postscripts.

  - *The Magic Numbers of Dr. Matrix* (Dorset, 1990).

- [UNEXPECTED], *The Unexpected Hanging and Other Mathematical Diversions* (Simon and Schuster, 1969).
    - *The Unexpected Hanging and Other Mathematical Diversions* (Simon and Schuster, 1986).
    - *The Unexpected Hanging and Other Mathematical Diversions* (University of Chicago Press, 1991).

    "With a new afterword and expanded bibliography."

    - *Knots and Borromean Rings, Rep-Tiles, and Eight Queens* (Cambridge University Press / Mathematical Association of America, 2014).

    Updated.

- [SIXTH], *Martin Gardner's Sixth Book of Mathematical Games from Scientific American* (W. H. Freeman, 1971).
  - *Sixth Book of Mathematical Games from Scientific American* (Charles Scribner's Sons, 1971).
  - *Sixth Book of Mathematical Diversions from Scientific American* (University of Chicago Press, 1983).

  "Phoenix edition" with slight title change.

- [CARNIVAL], *Mathematical Carnival* (Knopf, 1975).

  – *Mathematical Carnival* (Vintage, 1977).
  – *Mathematical Carnival* (Mathematical Association of America, 1989).

  Foreword by John Horton Conway, and "a new postscript and a new bibliography by Mr. Gardner."

1	Sprouts and Brussel Sprouts	3-11	Jul 67
2	Penny Puzzles	12-26	Feb 66
3	Aleph-null and Aleph-one	27-40	Mar 66
4	Hypercubes	41-54	Nov 66
5	Magic Stars and Polyhedrons	55-65	Dec 65
6	Calculating Prodigies	66-76	Apr 67
7	Tricks of Lightning Calculators	77-88	May 67
8	The Art of M. C. Escher	89-102	Apr 66
9	The Red-Faced Cube and Other Problems	103-122	Nov 65
10	Card Shuffles	123-138	Oct 66
11	Mrs. Perkins Quilt and Other Square-Packing Problems	139-149	Sep 66
12	The Numerology of Dr. Fleiss	150-160	Jul 66
13	Random Numbers	161-172	Jul 68
14	The Rising Hourglass and Other Physics Problems	173-193	Aug 66
15	Pascal's Triangle	194-207	Dec 66
16	Jam, Hot, and Other Games	208-225	Feb 67
17	Cooks and Quibble-Cooks	226-239	May 66
18	Piet Hein's Superellipse	240-254	Sep 65
19	How to Trisect and Angle	255-266	Jun 66
	Postscript [1989]	267-280	
	Bibliography	281-297	

- [MAGICSHOW], *Mathematical Magic Show* (Knopf, 1977).

  – *Mathematical Magic Show* (Vintage, 1978).
  – *Mathematical Magic Show* (Mathematical Association of America, 1989).

  Foreword by Ronald L. Graham and Persi Diaconis and "a new postscript and a new bibliography."

1	Nothing	15-28	Feb 75
2	More Ado about Nothing	29-34	NOT *SA*
3	Game Theory, Guess It, Foxholes	35-49	Dec 67

- [CIRCUS], *Mathematical Circus* (Knopf, 1980).

  - *Mathematical Circus* (Vintage, 1981).
  - *Mathematical Circus* (Mathematical Association of America, 1992).

    Preface by Donald Knuth and "a postscript from the author and a new bibliography."

- [WHEELS], *Wheels, Life, and Other Mathematical Amusements* (W. H. Freeman, 1983).

• [PENROSE], *Penrose Tiles and Trapdoor Ciphers* (W. H. Freeman, 1989).

Subtitled on jacket " ... and the Return of Dr. Matrix."

 — *Penrose Tiles and Trapdoor Ciphers and the Return of Dr. Matrix* (Mathematical Association of America, 1997).

   With new postscript (which is missing illustrations in the first printing.)

- [FRACTAL], *Fractal Music, Hypercards, and More: Mathematical Recreations from Scientific American Magazine* (W. H. Freeman, 1992).

- [LAST], *The Last Recreations: Hydras, Eggs, and Other Mathematical Mystifications* (Copernicus, 1997).

3.3   Other Books with Columns

1. *Mathematics: An Introduction to Its Spirit and Use* (W. H. Freeman, 1979). "Readings from *Scientific American*"; solutions found
   on pp. 235-241.

Topological Diversion, Including a Bottle with No Inside or Outside	Jul 63
On the Fabric of Inductive Logic, and Some Probability Paradoxes	Mar 76
Boolean Algebra, Venn Diagrams and the Propositional Calculus	Feb 69

2. Underwood Dudley (Ed), *Readings for Calculus* (Mathematical Association of America, 1993) 190-196. (MAA Notes, Vol. 31)

This starts with a lengthy introduction, by Dudley, on recreational mathematics. Two original problems (#5 and #7) are omitted here.

Nine More Problems	May 59

3. *Science: Good, Bad, and Bogus* (Prometheus, 1996).

This article was not collected in any other book.

Targ's ESP Teaching Machine	Oct 75

4. *The Night is Large: Collected Essays, 1938-1995* (St. Martins Press, 1996).

Can Time Stop? The Past Change?	Mar 79
The Laffer Curve	Dec 81
Mr. Apollinax Visits New York	May 61

5. Caroline Postelle Clotfelter (Ed), *On the Third Hand: Humor in the Dismal Science, an Anthology* (Univ of Michigan Press, 1996) 291-301.

The Laffer Curve	Dec 81

6. *My Best Mathematical and Logic Puzzles* (Dover, 1994).

The first 54 puzzles comprise all the puzzle columns that were collected in FIRST, SECOND, and NEW; however these original puzzles were excluded: "The Circle on the Chessboard," "The Cork Plug," "Professor on the Escalator," "Barr's Belt," "Two Pentomino Posers," and "A Pair of Digit Puzzles." The remaining 12 puzzles were drawn from *Games* magazine (January and November of 1978); four of these appeared in later columns: #55 in MAGICSHOW-5; #57 in UNEXPECTED-15; #58 in MAGICSHOW-15; #64 in TIME-16.

7. William Frucht (Ed), *Imaginary Numbers* (Wiley, 1999) 169-180.

   Church of the Fourth Dimension   Jan 62

8. Sharon Schwarze and Harvey Lape (Eds), *Thinking Socratically: Critical Thinking about Everyday Issues* (Prentice-Hall, 2001) 293-299.

   Fliess, Freud, and Biorhythm   Jul 66

9. *The Colossal Book of Mathematics* (Norton, 2001).

   Anthology of 50 *Scientific American* columns. While every chapter may have been lightly edited, some clearly have new addenda as noted.

   Arithmetic and Algebra
1	The Monkey and the Coconuts	Apr 58
2	The Calculus of Finite Differences	Aug 61
3	Palindromes: Words and Numbers	Aug 70

   Plane Geometry
4	Curves of Constant Width	Feb 63
5	Rep-Tiles †	May 63
6	Piet Hien's Superellipse †	Sep 65
7	Penrose Tiles ‡	Jan 77
8	The Wonders of a Planiverse	Jul 80

   Solid Geometry and Higher Dimensions
9	The Helix ‡	Jun 63
10	Packing Spheres †	May 60
11	Spheres and Hyperspheres ‡	May 68
12	The Church of the Fourth Dimension †	Jan 62
13	Hypercubes ‡	Nov 66
14	Non-Euclidean Geometry	Oct 81

   Symmetry
15	Rotations and Reflections ‡	May 62
16	The Amazing Creations of Scott Kim	Jun 81
17	The Art of M. C. Escher	Nov 65

Notes:
†This chapter contains a few paragraphs of new addenda.
‡This chapter contains many paragraphs of new addenda.

10. Richard Dawkins (Ed), *The Oxford Book of Modern Science Writing* (Oxford Univ Press, 2008) 276-284.

The Fantastic Combinations of John Conway's 'Life'   Oct 70

11. Editors of *Scientific American, Martin Gardner: The Magic and Mystery of Numbers* (Scientific American, 2014). e-book only

An electronic collection of 18 columns to coincide with the Centennial articles and related webpages.

3.4   Published Correspondence

1. *Scientific American* **196** (March 1957) 8.

Explains how to fold a heptahexaflexagon (cf. December 1956)

2. *Scientific American* **196** (April 1957) 14.

Responds to two letters about puzzles (cf. February 1957)

3. *Scientific American* **205** (August 1961) 12.

Explains why two white bishops are OK (cf. June 1961)

4. *Scientific American* **207** (July 1962) 12.

Responds to a loophole in a left-right experiment (cf. May 1962)

5. *Scientific American* **212** (March 1965) 11.

Caltrop is defined by a reader (cf. February 1965)

6. *Scientific American* **214** (January 1966) 8-9.

Explains why Martian canals are illusions (cf. August 1965)

7. *Scientific American* **223** (December 1970) 6.

Reader argues that some animals use rotary locomotion (cf. September 1970)

8. *Scientific American* **246** (March 1982) 8.

Reply to comments on the Neo-Laffer curve (cf. December 1981)

3.5 Original Column Index

Original column		Book appearance	
Month	Year	Book	Chapter
December	1956	FIRST	1 †
January	1957	FIRST	2
February	1957	FIRST	3
March	1957	FIRST	4
April	1957	FIRST	5
May	1957	FIRST	6
June	1957	FIRST	7
July	1957	FIRST	8
August	1957	FIRST	9
September	1957	FIRST	10
October	1957	FIRST	11
November	1957	FIRST	12
December	1957	FIRST	13
January	1958	FIRST	14
February	1958	FIRST	15
March	1958	FIRST	16
April	1958	SECOND	9 †
May	1958	SECOND	2
June	1958	SECOND	3
July	1958	SECOND	4
August	1958	SECOND	5
September	1958	SECOND	6 †
October	1958	SECOND	7
November	1958	SECOND	17
December	1958	SECOND	1
January	1959	SECOND	10
February	1959	SECOND	11
March	1959	SECOND	12
April	1959	SECOND	13
May	1959	SECOND	14
June	1959	SECOND	15
July	1959	SECOND	16
August	1959	SECOND	8
September	1959	SECOND	18
October	1959	SECOND	19 †
November	1959	NEW	14
December	1959	NEW	2

Original column		Book appearance	
Month	Year	Book	Chapter
January	1960	DRMATRIX	1*
February	1960	NEW	3
March	1960	NEW	4
April	1960	NEW	6
May	1960	NEW	7 †
June	1960	NEW	5
July	1960	NEW	8
August	1960	NEW	9
September	1960	NEW	10
October	1960	NEW	12
November	1960	NEW	13
December	1960	NEW	1
January	1961	DRMATRIX	2
February	1961	NEW	15
March	1961	NEW	16
April	1961	NEW	17
May	1961	NEW	11
June	1961	NEW	19
July	1961	NEW	18
August	1961	NEW	20 †
September	1961	UNEXPECTED	2
October	1961	UNEXPECTED	3
November	1961	UNEXPECTED	4
December	1961	UNEXPECTED	5
January	1962	UNEXPECTED	6
February	1962	UNEXPECTED	7 †
March	1962	UNEXPECTED	8 †
April	1962	UNEXPECTED	9
May	1962	UNEXPECTED	10 †
June	1962	UNEXPECTED	11
July	1962	UNEXPECTED	12
August	1962	UNEXPECTED	13
September	1962	UNEXPECTED	14
October	1962	UNEXPECTED	15
November	1962	UNEXPECTED	16
December	1962	UNEXPECTED	17

| Original column | | Book appearance | |
Month	Year	Book	Chapter
January	1963	DR MATRIX	3
February	1963	UNEXPECTED	18 †
March	1963	UNEXPECTED	1 †
April	1963	UNEXPECTED	20
May	1963	UNEXPECTED	19 †
June	1963	SIXTH	1 †
July	1963	SIXTH	2 †
August	1963	SIXTH	3
September	1963	SIXTH	4
October	1963	SIXTH	5
November	1963	SIXTH	6
December	1963	SIXTH	8
January	1964	DR MATRIX	5
February	1964	SIXTH	7
March	1964	SIXTH	9
April	1964	SIXTH	10
May	1964	SIXTH	11
June	1964	SIXTH	12
July	1964	SIXTH	13
August	1964	SIXTH	14
September	1964	SIXTH	15
October	1964	SIXTH	16
November	1964	SIXTH	17
December	1964	SIXTH	18
January	1965	DR MATRIX	6
February	1965	SIXTH	19
March	1965	SIXTH	20
April	1965	SIXTH	22 †
May	1965	SIXTH	21
June	1965	SIXTH	23
July	1965	SIXTH	24
August	1965	SIXTH	25
September	1965	CARNIVAL	18 †
October	1965	MAGIC SHOW	13
November	1965	CARNIVAL	9
December	1965	CARNIVAL	5

| Original column | | Book appearance | |
Month	Year	Book	Chapter
January	1966	DR MATRIX	7
February	1966	CARNIVAL	2
March	1966	CARNIVAL	3 †
April	1966	CARNIVAL	8 †
May	1966	CARNIVAL	17
June	1966	CARNIVAL	19
July	1966	CARNIVAL	12
August	1966	CARNIVAL	14
September	1966	CARNIVAL	11
October	1966	CARNIVAL	10
November	1966	CARNIVAL	4 †
December	1966	CARNIVAL	15
January	1967	DR MATRIX	9
February	1967	CARNIVAL	16
March	1967	MAGIC SHOW	15
April	1967	CARNIVAL	6
May	1967	CARNIVAL	7
June	1967	MAGIC SHOW	11
July	1967	CARNIVAL	1 †
August	1967	MAGIC SHOW	4
September	1967	MAGIC SHOW	6
October	1967	MAGIC SHOW	14
November	1967	MAGIC SHOW	5
December	1967	MAGIC SHOW	3
January	1968	DR MATRIX	10
February	1968	MAGIC SHOW	17
March	1968	MAGIC SHOW	12
April	1968	CIRCUS	20
May	1968	CIRCUS	3 †
June	1968	MAGIC SHOW	7
July	1968	CARNIVAL	13
August	1968	MAGIC SHOW	10
September	1968	MAGIC SHOW	8
October	1968	MAGIC SHOW	16
November	1968	MAGIC SHOW	18
December	1968	MAGIC SHOW	9

Original column		Book appearance	
Month	Year	Book	Chapter
January	1969	DR MATRIX	12
February	1969	CIRCUS	8
March	1969	CIRCUS	13
April	1969	CIRCUS	15
May	1969	CIRCUS	6
June	1969	CIRCUS	7
July	1969	CIRCUS	2
August	1969	CIRCUS	14 †
September	1969	CIRCUS	17
October	1969	DR MATRIX	13
November	1969	CIRCUS	4
December	1969	CIRCUS	12
January	1970	CIRCUS	18
February	1970	CIRCUS	11
March	1970	CIRCUS	10
April	1970	CIRCUS	16
May	1970	CIRCUS	1
June	1970	CIRCUS	5
July	1970	WHEELS	2
August	1970	CIRCUS	19 †
September	1970	WHEELS	1
October	1970	WHEELS	20 †
November	1970	WHEELS	3
December	1970	WHEELS	5 †
January	1971	DR MATRIX	14
February	1971	WHEELS	21
March	1971	WHEELS	4 †
April	1971	WHEELS	6
May	1971	WHEELS	7 †
June	1971	CIRCUS	9
July	1971	WHEELS	8
August	1971	WHEELS	9
September	1971	WHEELS	10
October	1971	WHEELS	11
November	1971	WHEELS	12
December	1971	WHEELS	13

Original column		Book appearance	
Month	Year	Book	Chapter
January	1972	WHEELS	14
February	1972	DRMATRIX	15
March	1972	WHEELS	15
April	1972	WHEELS	16
May	1972	WHEELS	17
June	1972	WHEELS	18
July	1972	WHEELS	19
August	1972	KNOTTED	2
September	1972	KNOTTED	3
October	1972	KNOTTED	1
November	1972	KNOTTED	4
December	1972	KNOTTED	5 †
January	1973	KNOTTED	9
February	1973	KNOTTED	10
March	1973	KNOTTED	7
April	1973	KNOTTED	8
May	1973	KNOTTED	6
June	1973	KNOTTED	11
July	1973	KNOTTED	13 †
August	1973	DRMATRIX	16
September	1973	KNOTTED	12
October	1973	KNOTTED	16
November	1973	KNOTTED	17
December	1973	KNOTTED	18
January	1974	KNOTTED	20
February	1974	KNOTTED	19
March	1974	KNOTTED	14
April	1974	KNOTTED	15
May	1974	TIME	1 †
June	1974	DRMATRIX	17
July	1974	TIME	2
August	1974	TIME	3
September	1974	TIME	4
October	1974	TIME	5 †
November	1974	TIME	6
December	1974	TIME	7 †

| Original column | | Book appearance | |
Month	Year	Book	Chapter
January	1975	TIME	8
February	1975	MAGICSHOW	1 †
March	1975	TIME	9
April	1975	TIME	10 †
May	1975	TIME	11
June	1975	TIME	12
July	1975	TIME	13
August	1975	TIME	14
September	1975	DRMATRIX	18
October	1975	**	**
November	1975	TIME	15
December	1975	TIME	16
January	1976	TIME	17
February	1976	TIME	18
March	1976	TIME	19 †
April	1976	LAST	23
May	1976	MAGICSHOW	19 †
June	1976	TIME	20
July	1976	TIME	21
August	1976	TIME	22
September	1976	PENROSE	4 †
October	1976	PENROSE	5
November	1976	DRMATRIX	19
December	1976	PENROSE	3
January	1977	PENROSE	1 †
February	1977	PENROSE	6
March	1977	PENROSE	8
April	1977	PENROSE	9
May	1977	PENROSE	10
June	1977	PENROSE	11
July	1977	PENROSE	12
August	1977	PENROSE	13
September	1977	PENROSE	15
October	1977	PENROSE	16 †
November	1977	PENROSE	17 †
December	1977	DRMATRIX	20

Original column		Book appearance	
Month	Year	Book	Chapter
January	1978	PENROSE	18
February	1978	PENROSE	19
March	1978	PENROSE	20
April	1978	FRACTAL	1 †
May	1978	FRACTAL	2
June	1978	FRACTAL	3 †
July	1978	FRACTAL	4
August	1978	FRACTAL	5
September	1978	FRACTAL	6
October	1978	FRACTAL	7
November	1978	FRACTAL	8
December	1978	DRMATRIX	21
January	1979	FRACTAL	10
February	1979	FRACTAL	11
March	1979	FRACTAL	12 †
April	1979	FRACTAL	13 †
May	1979	FRACTAL	14
June	1979	FRACTAL	15
July	1979	FRACTAL	16 †
August	1979	FRACTAL	17
September	1979	FRACTAL	18
October	1979	FRACTAL	20
November	1979	FRACTAL	21
December	1979	LAST	21
January	1980	LAST	13
February	1980	LAST	6
March	1980	LAST	7
April	1980	LAST	3
May	1980	LAST	8
June	1980	LAST	9
July	1980	LAST	1 †
August	1980	LAST	11
September	1980	DRMATRIX	22
October	1980	LAST	20
November	1980	LAST	10
December	1980	LAST	12

| Original column | | Book appearance | |
Month	Year	Book	Chapter
February	1981	LAST	15
April	1981	LAST	16
June	1981	LAST	17 †
August	1981	LAST	18
October	1981	LAST	19 †
December	1981	KNOTTED	21
August	1983	LAST	2 †
September	1983	LAST	5 †
June	1986	LAST	22

Notes:

† This also appeared in *The Colossal Book of Mathematics* (2001), possibly with new addenda.

* This is also chapter 20 of SECOND.

** This article on ESP was not collected in these books; it is chapter seven of *Science: Good, Bad, and Bogus* (Prometheus, 1981) 75-90.

3.6 Index of the Puzzle Columns

Nearly every year one of the *Scientific American* columns was a collection of puzzles, rather than an essay. These puzzles were always anticipated and popular. Listing them in detail is difficult since they are not easily categorized in a way that they can appear in an index (though many are listed in the Appendices C and D). Therefore the puzzles are all listed here with their title and a one-line description.

All of these puzzles were collected and numbered in the book *The Colossal Book of Short Puzzles and Problems* (2006). After each puzzle's title the puzzle's number in that book is shown in brackets.

February 1957 — FIRST, Chapter 3

The Returning Explorer [6.6]
    Variant of the polar bear puzzle.
Draw Poker [11.4]
    What five cards guarantee a win at poker?

The Mutilated Chessboard [1.5]
  Domino tiling with opposite corners removed.
The Fork in the Road [13.22]
  Truth-teller/liar problem.
Scrambled Box Tops [9.12]
  Mislabelled boxes with two colored marbles.
Bronx vs. Brooklyn [2.7]
  Unequal likelihood of getting an uptown train.
Cutting the Cube [6.7]
  Minimum number of cuts to cut a cube into 27 cubes.
The Early Commute [4.22]
  How long did the commuter walk before being picked up?
Counterfeit Coins [9.15]
  One weighing on a pointer scale.

November 1957 — FIRST, Chapter 12

The Touching Cigarettes [16.14]
  Place six or more cigarettes so that they touch each other.
Two Ferryboats [4.23]
  How wide is the river given where they cross?
Guess the Diagonal [5.12]
  Length of a rectangle inscribed in a quarter circle.
The Efficient Electrician [9.19]
  Identify the ends of wires using short circuits.
Cross the Network [8.7]
  Enter every room puzzle on a torus.
The Twelve Matches [5.16]
  Use twelve matches to enclose four square units.
Hole in the Sphere [6.14]
  Volume of remaining sphere with a six inch hole in it.
The Amorous Bugs [5.22]
  How far do the bugs go to the center of the square?
How Many Children [13.13]
  Give the number of children given very little information.

August 1958 — SECOND, Chapter 5

The Twiddled Bolts [16.12]
  Do the two bolts move in or out?
The Flight around the World [10.9]
  Schedule the refuelings.
The Circle on the Chessboard [5.13]
  Largest circle entirely on black squares.

The Cork Plug [6.8]
  Solid with square, triangular and circular projections.
The Repetitious Number [3.12]
  Self-working number prediction: divide by 7, 11, and 13.
The Colliding Missiles [4.11]
  How far apart are they one minute before they collide?
The Sliding Pennies [10.7]
  Transform a triangle of six pennies into a hexagon.
Handshakes and Networks [1.16]
  The number of men shaking an odd number of hands is even.
The Triangular Duel [9.30]
  Three "duelists" with unequal probabilities of success.

May 1959 — SECOND, Chapter 14

Crossing the Desert [10.10]
  Schedule the refuelings.
The Two Children [2.6]
  If there is least one boy what is the probability of two boys?
Lord Dunsany's Chess Problem [12.5]
  White to mate in four moves; White has no pawns.
Professor on the Escalator [4.24]
  How long is an escalator given rates going up and down?
The Lonesome 8 [14.2]
  A cryptarithm puzzle; all X's and one 8.
Dividing the Cake [9.11]
  Divide a cake fairly among n people.
The Folded Sheet [1.9]
  Fold a numbered 2 × 4 sheet into numerical order.
The Absent-Minded Teller [4.14]
  The dollars and cents amounts are switched.
Water and Wine [9.21]
  Relative proportions after mixing; also repeated mixings.

February 1960 — NEW, Chapter 3

Acute Dissection [7.6]
  Dissect an obtuse triangle into acute triangles.
How Long is a Lunar? [6.9]
  ... given that the moons surface equals its volume.
The Game of Googol [11.8]
  Optimal stopping; pick the slip with the largest number.
Marching Cadets and Trotting Dogs [4.25]
  How far does a dog travel around a moving formation?

Barr's Belt [8.8]
  Fold a long strip of paper into uniform thickness.
White, Black and Brown [13.15]
  "What is the color of Professor Black's hair?"
The Plane in the Wind [4.16]
  The effect of wind direction on round-trip times.
What Price Pets? [4.15]
  Diophantine analysis given various costs.

October 1960 — NEW, Chapter 12

The Game of Hip [11.12]
  Color a 6 × 6 board, so no square has monochromatic corners.
A Switching Puzzle [10.13]
  A shunting problem; switch two cars with a thin tunnel.
Beer Signs on the Highway [4.19]
  The number of signs per minute times 10 is the car's speed.
The Sliced Cube and the Sliced Doughnut [6.10]
  Hexagonal slice of cube; intersecting circles slice of doughnut.
Bisecting Yin and Yang [5.14]
  Draw line that bisects both the black and the white regions.
The Blue-Eyed Sisters [2.8]
  How many sisters if even chance two are both blue-eyed?
How Old is the Rose-Red City [4.13]
  Arithmetic puzzle : "A rose-red city half as old as Time."
Tricky Track [13.17]
  Given minimal data: "Which school won the high jump?"
Termite and the 27 Cubes [1.6]
  Can a termite, going to the center, bore through every cube?

June 1961 — NEW, Chapter 19

Collating the Coins [10.6]
  Rearrange five coins, with a penny that is adjacent to a dime.
Time the Toast [10.22]
  Scheduling an old-fashioned toaster.
Two Pentomino Posers [7.10]
  Cut a tiling and reassemble; tile with no interior pieces.
A Fixed-Point Theorem [4.3]
  Does the monk occupy the same point a day later?
A Pair of Digit Puzzles [14.8, 14.9]
  $OODDF = \sqrt{WONDERFUL}$; sum with trace of rook moves.
How Did Kant Set His Clock? [9.13]
  Resetting a stopped clock after a walk.

Playing Twenty Questions with Known Probability Values [9.29]
  Guess a card in a biased deck (solved with the Huffman-code).
Don't Mate in One [12.3]
  In a given chess position, how can white move and not mate?
Find the Hexahedrons [6.11]
  Find the (seven) convex hexahedrons.

February 1962 — UNEXPECTED, Chapter 7

A Digit-Placing Problem [1.8]
  Place digits at nodes with no consecutive digits connected.
The Lady or the Tiger? [2.9]
  What is the probability of picking the lady, given conditions?
A Tennis Match [13.18]
  Who served first given win-loss records?
The Colored Bowling Pins [1.14]
  Two-color so there is no monochromatic equilateral triangle.
The Problem of Six Matches [8.6]
  Enumerate all possible planar graphs with unit-length edges.
Two Chess Problems: Minimum and Maximum Attack [12.6]
  Eight pieces to attack the fewest squares / the most squares.
How Far Did the Smiths Travel? [4.8]
  Distance calculation given time and distance data.
Predicting a Finger Count [3.11]
  Counting back and forth on fingers; where do you end up?

October 1962 — UNEXPECTED, Chapter 15

The Seven File Cards [7.5]
  How much of a standard piece of paper can be covered?
A Blue-Empty Graph [1.17]
  Ramsey problem; six people have monochromatic triangle.
Two Games in a Row [2.10]
  ... are avoided when alternately against unequal players.
A Pair of Cryptarithms [14.3]
  Given the parity of each digit; given that each digit is prime.
Dissecting a Square [7.1]
  Dissect a square into five equal pieces.
Traffic Flow in Floyd's Knob [10.14]
  A traffic maze given a map with direction restrictions.
Littlewood's Footnotes [15.10]
  Avoiding an infinite regress when translating footnotes.
Nine to One Equals 100 [3.13]
  Introduce +'s and −'s to 123456789 to produce 100.

The Crossed Cylinders [6.13]
Calculate the volume of the intersection of two cylinders.

November 1963 — SIXTH, Chapter 6

The Rigid Square [5.27]
Using unit-length rods construct a rigid square in the plane.
A Penny Bet [2.12]
Flip two pennies vs one; calculate the odds.
Three-Dimensional Maze [10.15]
Navigate a given $4 \times 4 \times 4$ maze.
Gold Links [9.18]
Minimum cuts of a chain of 23 links for the amounts 1 to 23.
Word Squares [15.12]
Construct a word square on "circle".
The Three Watch Hands [4.17]
When are the three hands "close" together?
Three Cryptarithms [14.4]
A normal one; two with each digit used twice.
Maximizing Chess Moves [12.7]
Eight pieces to maximize the number possible moves.
Folding a Möbius Strip [8.9]
Smallest aspect ratio to allow the folding.

June 1964 — SIXTH, Chapter 12

The Trip around the Moon [10.11]
Circular refueling schedule.
The Rectangle and the Oil Well [6.16]
Missing measurement geometry problem.
Wild Ticktacktoe [11.5]
Either player can play X or O.
Coins of the Realm [1.11]
Minimum set so that 1 to 100 can be done with 1 or 2 coins.
Bills and Two Hats [2.13]
Dividing a set of bills to maximize the expected draw.
Dudeney's Word Square [15.13]
Word square with words in any direction.
Ranking Weights [9.17]
Ranking/sorting weights with a balance scale.
Queen's Tours [12.18]
Five extremal tasks.

March 1965 — SIXTH, Chapter 20

Coleridge's Apples [4.12]
"Give half of what remains and half an apple."
Reversed Trousers [8.11]
Tie ankles together and then remove and reverse pants.
Coin Game [11.7]
Pursuit game on a graph; a win in six moves?
Truthers, Liars, and Randomizers [13.23]
Ask three questions to identify one of each.
Gear Paradox [16.11]
Explain how the illustrated geared device works.
Form a Swastika [16.7]
...using cigarettes and sugar cubes.
Blades of Grass Game [11.11]
Connect the ends of six lines; odds there is a loop?
Casey at the Bat [11.3]
Least possible runs if Casey came to bat every inning.
The Eight-Block Puzzle [10.17]
Minimum number of moves for a 3 × 3 sliding block puzzle.

November 1965 — CARNIVAL, Chapter 9

The Red-Faced Cube [10.18]
Roll a cube with a red face on chessboard.
The Three Cards [13.19]
Given position constraints name the three cards.
The Key and the Keyhole [8.5]
Topological; move the key from one loop to another.
The Anagram Dictionary [15.14]
Seven questions: 'What is the first entry starting with B?'
A Million Points [9.9]
Find a line that evenly divides points in a closed curve.
Lady on the Lake [9.26]
Row to the shore of circular lake to avoid pursuer on shore.
Killing Squares and Rectangles [1.20]
Remove matches so that no squares/rectangles remain.
Cocircular Points [5.15]
Given circles and rectangles, find cocircular points.
The Poisoned Glass [9.23]
Fewest poison tests of liquid taken from many glasses.

August 1966 — CARNIVAL, Chapter 14
[All physics puzzles.]

Two Hundred Pigeons [17.1]
   Weight of pigeons flying inside sealed truck.
The Rising Hourglass [17.2]
   Why does the hourglass wait to rise in a cylinder of water?
Iron Torus [17.3]
   Does the hole expand when it is heated?
The Suspended Horseshoe [17.4]
   Pick up a cardboard horshoe and toothpick with another one.
Center the Cork [17.5]
   Make a cork float in the middle of a glass.
Oil and Vinegar [17.6]
   How to pour out a desired proportion from a single bottle.
Carroll's Carriage [17.7]
   Mechanics of a carriage with elliptical wheels.
Magnet Testing [17.8]
   Give two bars, how to determine which one is a magnet?
Melting Ice Cube [17.9]
   Level of water when ice melts, near the freezing point.
Stealing Bell Ropes [17.10]
   How to get the most bell rope by climbing with a knife.
Moving Shadow [17.11]
   Does the top of a shadow move at a uniform speed over time?
The Coiled Hose [17.12]
   Why doesn't water flow naturally through a coiled hose?
Egg in a Bottle [17.13]
   How to get a egg out of a bottle.
Bathtub Boat [17.14]
   Will a boat rise if its cargo is jettisoned and the cargo sinks?
Ballon in Car [17.15]
   How does a ballon move when a car accelerates?
Hollow Moon [17.16]
   Gravity inside a hollow moon?
Lunar Bird [17.17]
   Would a bird fly faster on the moon, with less gravity?
The Compton Tube [17.18]
   About a toroidal glass tube of liquid that flips.
Fishy Problem [17.19]
   Weight of a fish in a bowl when you hold its tail.
Bicycle Paradox [17.20]
   What happens when you pull back on the lower pedal?

Inertial Drive [17.21]

    Can you move a boat by jerking on a rope from within?

Worth of Gold [17.22]

    "What is worth more, a pound of \$10 gold pieces or ... "

Switching Paradox [17.23]

    Explain the wiring trick, given the switches and bulbs.

February 1967 — MAGICSHOW, Chapter 15

Interrupted Bridge Game [9.8]

    How to finish dealing hands if interrupted?

Nora L. Aron [14.5]

    Cryptarithm: NORA $\times$ L = ARON.

Polyomino Four-Color Problem [8.15]

    Two problems about tilings that need four colors.

How many Spots? [5.21]

    Place spots that are not within $\sqrt{2}$ of each other.

The Three Coins [2.15]

    Unseen coins are flipped; best strategy to get all heads?

The 25 Knights [12.4]

    Can knights on every square of a 5×5 board move at once?

The Dragon Curve [9.28]

    Explain different ways to define the dragon curve.

The Ten Soldiers [1.15]

    Longest ascending or descending subsequence.

A Curious Set of Integers [3.6]

    Any two form a product that is $n^2 - 1$; $\{1,3,8,120,\text{"?"}\}$.

November 1967 — MAGICSHOW, Chapter 5

The Cocktail Cherry [9.7]

    Remove the cherry by moving two matches.

The Papered Cube [6.12]

    Largest cube that can be covered by a square sheet.

Lunch at the TL Club [13.24]

    Truth-tellers and liars seated around a table.

A Fair Division [9.16]

    Not enough info? Based on parity of the tens digit of n^2.

Tri-Hex [11.10]

    Ticktacktoe on a graph different from the 3×3 grid.

Langford's Problem [1.13]
 Arrange $2n$ pairs of digits so there are i digits between i and i.
Overlap Squares [5.18]
 A 4 inch square has a corner at the center of a 3 inch square.
Families in Fertilia [2.14]
 If no children after first boy, what is the expected family size?
Christmas and Halloween [3.9]
 Prove that Oct. 31 = Dec. 25.
Knot the Rope [8.10]
 Can you tie a knot in a rope connected to the wrists?

August 1968 — MAGICSHOW, **Chapter 10**
 ["Ridiculous Questions"]

How many earrings are worn by the villagers? [4.9]
Can a solid be unstable on every face? [6.1]
What is the missing (ternary) number? [3.5]
Find three errors in the five equations. [13.6]
Why pick the poorly trimmed barber? [13.4]
Ticktacktoe with ten squares, four on the bottom. [11.1]
Probability that three out four envelopes correct. [2.2]
Are two corners and center of an irregular solid coplanar? [6.3]
Joke 4×4 crossword; ems, eyes, els, ease = mole's. [15.4]
Probability that three points are in the same hemisphere. [2.4]
"...how many apples would you have?" [4.1]
Perimeter of a triangle formed by three tangents. [5.10]
"Twelve thousand, twelve hundred and twelve." [3.1]
"Chemical reaction took 80 minutes" with a jacket." [9.1]
Cutting a genus two Möbius-like band. [8.3]
Relative areas of polygons with equal perimeters. [5.4]
Can you build a 6×6 cube from 1×4 blocks? [6.4]
Dead fly in the coffee. [9.2]
What is the area of the given figure? [5.9]
"This parrot will repeat every word it hears." [13.2]
Where does the trisector of a square's corner intersect? [5.5]
Can two spheres go through pipe in opposite directions? [10.1]
Three ways a barometer can be used to determine height. [9.3]
Make a cube with five unbroken matches. [6.5]
Relative odds of partners having all the clubs vs no clubs. [2.1]
The shortchanging bellboy. [9.5]

April 1969 — CIRCUS, Chapter 15

Rotating Round Table [1.23]
 Will a cyclic permutation of any derangement have two hits?
Single-Check Chess [12.14]
 Can white force the first check of the game in five moves?
Word-Guessing Game [11.6]
 Jotto-like game except that only the parity of hits is reported.
Triple Beer Rings [5.19]
 Calculate the area of intersection of three overlapping circles.
Two-Cube Calendar [1.4]
 Generate 01, 02, ..., 31 from twelve digits on two cubes.
Uncrossed Knight's Tours [12.16]
 Find a 17 move tour on order 6 board which is noncrossing.
Two [One] Urn Problems [2.16]
 Given a removal strategy, what are the odds black remains?
Ten Quickies
 Time 15 minutes using 7 and 11 minute hourglasses. [10.2]
 Minimum wear when rotating tires. [9.4]
 Probability of top cards being same color after a cut. [2.5]
 For what nondecimal base is 121 a perfect square. [3.3]
 Form eight equilateral triangles with six unit line. [5.2]
 Angle trisection and dividing 2^i by 3. [5.8]
 "How many horses ... if you call the cows horses?" [4.5]
 Translate "He spoke from 2222222222222 people." [15.6]
 Age of someone born BC and dying AD. [4.2]
 Two queries to tell if truth-teller, liar or alternater. [13.11]

February 1970 — CIRCUS, Chapter 11

Eccentric Chess [12.8]
 Remove knights in five moves; a non-chess counting paradox.
Talkative Eve [14.6]
 A cryptarithm: EVE / DID = .TALKTALK...
Three Squares [5.23]
 Using geometry alone prove $\angle A + \angle B = \angle C$.
Pohl's Proposition [1.10]
 Binary stunt to count flips of a set of coins.
Escott's Sliding Blocks [10.20]
 A difficult puzzle with irregularly shaped blocks.
Red, White, and Blue Weights [9.22]
 Separate three pairs of unequal weights in two weighings.
The 10-Digit-Number [1.12]
 The ith digit is the number of times i is a digit in the number.

Bowling-Ball Pennies [1.19]
   Remove the fewest pennies to destroy all equilateral triangles.
Knockout Geography [11.2]
   Who wins when playing Geography with the 50 states?

November 1970 — WHEELS, Chapter 3

The Knotted Molecule [8.16]
   Shortest rectlinear 3-D knotted chain which bends each unit.
Pied Numbers [3.17]
   Express integers ≤ 20 using fewest π's floor function allowed.
The Five Congruent Polygons [7.3]
   Cut the figure into five congruent polygons.
Starting a Chess Game [12.9]
   Permute the 32 pieces always moving to an empty square.
The Twenty Bank Deposits [4.20]
   Which generalized Fibonacci sequence leads to 1,000,000?
The First Black Ace [2.17]
   The expected location of the first black ace in a deck.
A Docecahedron-Quintomino Puzzle [1.21]
   5-color the edges so the faces have a unique cyclic permutation.
Scrambled Quotation [15.15]
   Reconstruct a phrase given all the first letters, all the second ...
The Blank Column [1.6]
   Will repeatedly typing the same thing give a blank column?
The Child with the Wart [13.20]
   Three ages; product is 36; sum is house number, oldest is ...

July 1971 — WHEELS, Chapter 8 ["Quickies"]

Build a wire cube use fewest solder joints. [6.2]
What proverb for a horse ignorant of cartesian coordinates? [15.2]
One knight and one opposite king; how long until check? [12.1]
Highest constant for a magic square made of playing cards. [1.3]
What predicate $P(n)$ is only true for $n < 1,000,000$? [13.9]
Why does barber prefers two Frenchman to one German. [13.3]
Number of proper intersections of two closed curves is even. [8.1]
Place a symbol between 2 and 3 so that number $2 < x < 3$. [3.10]
Ratio of distance to sixth floor and to the third floor. [3.2]
Maximum area of a isoceles triangle with sides: 1,1,x. [5.1]
Give three positive integers with a sum equal to their product. [4.6]
What is probability a string in trefoil pattern is knotted? [8.4]
If AB, BC, CD, and DE are common words what is DCABE? [15.8]

Show that $1324^n + 731^n = 1961^n$ is not true for any $n > 2$. [4.8]
What word begins and ends with "und"? [15.5]
What is the probability one sees three sides of the pentagon? [2.3]
"Make 11030 a person by adding 2 straight line segments." [15.7]
"...If at least one of them is lying, who is which?" [13.8]
Place nonattacking superqueens (with knight moves). [12.2]
ABCD + DCBA + XXXX = 12300, Xs are a permutation. [14.1]
Primes on a diagonal of the "primeval snake"? [3.8]
Positive integers such that $GCD(x,y)LCM(x,y) = xy$. [4.4]
"If only one is true, how many books does Feemster own?" [13.7]
How many month-day combinations are ambiguous? [1.1]
Why aren't manholes square? [16.1]
Count 10-digit numbers with distinct digits (no leading 0) [1.2]
"Is it possible that 72 hours later [it] was sunny?" [13.1]
Interpret "$2B \vee \sim 2B =$?" [15.1]
Relative areas of inscribed and circumscribed hexagons. [5.6]
"I was n years old in the year n^2", where $n < 1971$. [4.10]
I can write down the base (> 2) you are thinking of. [3.3]
What was the name of the Secretary General 35 years ago? [13.5]
How many white cubes can touch surfaces with a red cube? [6.17]
Rearrange four consecutive letters to spell a common word. [15.9]
Reach an island in the middle of a lake with a length of rope. [9.6]
Which way will the dog be facing after running back and forth? [4.7]

April 1972 — WHEELS, Chapter 16

The Flexible Band [8.13]
    Flex a band to pass through hole in itself while one end held.
The Rotating Disk [13.21]
    Points from by spinning; "What was each player's final score?"
Frieze Patterns [1.22]
    What is the rule that generates frieze patterns of numbers?
The Can of Beer [16.13]
    The height of the center of gravity of a non-full cyclinder.
The Three Coins [2.18]
    Three increasing value coins; A has 1; B has 2; points for heads.
Kobon Triangles [7.18]
    Maximum number of disjoint triangles with n line segments.
A Nine-Digit Problem [3.14]
    Use digits to create two products, both equal and maximal.
Crowning the Checkers [10.4]
    Start with n in a row and form kings; jump one, then two, ...

Charles Addams' Skier [16.8]
> Give six explanations for parallel tracks around a tree.

May 1973 — KNOTTED, **Chapter 6**

The Tour of the Arrows [10.6]
> Maximum length rook tour on order 4 board.

Five Couples [1.18]
> If everyone else shook a unique number of hands.

Square-Triangle Polygons [7.9]
> Convex polygons formed by unit squares and triangles.

Ten Statements [13.16]
> ... "Exactly ten statements on this list are false."

Pentomino Farms [7.11]
> Four maximum area problems using the twelve pentominoes.

The Uneven Floor [9.25]
> On a wavy floor can we put a table with four legs touching.

The Chicken-Wire Trick [16.10]
> How to put a pattern of chicken-wire creases into paper.

Where Was the King? [12.11]
> Where was the king before it was removed (retrograde chess).

Polypowers [3.15]
> Three problems like: $x^{x^x} = ((x^x)^x)^x$.

April 1974 — KNOTTED, **Chapter 15**

The Gunport Problem [7.7]
> Maximum number of unit holes, dominoes on $m \times n$ field.

Figures Never Lie [3.16]
> When does comedy routine of miscalculation work?

Functional Fixedness [16.9]
> Two creative problem solving tasks (e.g. tie two strings).

Monochromatic Chess [12.12]
> Color of the pawn if no piece moved from one color to another?

The Two Bookcases [10.8]
> Reverse two rods in a rectangle by only pivoting.

Irrational Probabilities [2.19]
> Using coin to choose between alternatives with irrational odds.

Who's Behind the Mad Hatter? [15.11]
> Who was in the dialogue? (It was the Jack of Spades).

Reverse the Fish [16.6]
> Move three matches to reverse the direction of the fish.

The Intersecting Circles [5.24]
> Show intersections of 3 unit circles, meeting a point, cocircular.

March 1975 — Time, Chapter 9

The Rubber Rope [16.15]
    Does a slow worm every cross an ever-stretching rope?
The Sigil of Scoteia [15.17]
    Can you read the secret alphabet used by James Cabell?
Integer Choice Game [11.13]
    Players pick x and y; smallest is best unless $x - 1 = y$.
Three Circles [5.25]
    Show colinearity of tangent intersections using 3 dimensions.
The Mutilated Score Sheet [12.13]
    Reconstruct a four move chess game from some of the moves.
Self-Numbers [3.18]
    $x \neq f(y)$ for any y, $f(y) = y + ($ the sum of the digits of $y)$.
The Colored Poker Chips [8.14]
    Minimum number of touching chips that require 4 colors.
Rolling Cubes [10.19]
    Invert all cubes in a 3×3 tray, with a vacancy in the middle.

November 1975 — Time, Chapter 16

What Symbol Comes Next [13.12]
    What is the sixth symbol in the sequence?
Which Symbol is Different [13.14]
    "Pick out the one that is 'most different'."
Cutting the Cake [9.20]
    Cut a square cake so the volume and frosting is equitable?
Two Cryptarithms [14.7]
    VINGT+CINQ+CINQ=TRENTE; EIN+EIN+EIN+EIN=VIER.
Lewis Carroll's "Sonnet" [15.16]
    A Carrollian poem in which pairs of words can be anagrammed.
Third-Man Theme [12.15]
    Given a two-king position add a piece so no legal play.

April 1977 — Penrose, Chapter 9

Pool-Ball Triangles [1.24]
    Arrange the 15 pool balls into a difference triangle.
Toroidal Cannibalism [8.17]
    Two linked toruses, one with a hole, move one is inside.

Exploring Tetrads [7.8]
    Give a 3-D polycube that's replicated and arranged to touch.
Knights and Knaves [13.25]
    Four short Smullyan liar/truth-teller problems.
Lost King's Tour [12.17]
    Two problems with a king that always changes directions.
Steiner Ellipses [5.17]
    Circumscribed and inscribed ellipses for 3-4-5 triangle.
Different Distances [5.26]
    Place seven counters on a 7 × 7 board, all distances distinct.
A Limerick Paradox [15.18]
    Limericks of decreasing length.

February 1979 — Fractal, Chapter 11

The Rotating Table [9.27]
    Repeatedly pick two corners and (possibly) invert glasses.
Turnablock [11.14]
    Game with two-color pieces, invert rectangular regions.
Persistence of Numbers [3.19]
    Smallest number that persists five steps of multiplying digits.
Nevermore [15.19]
    What rules were used for these three parodies of "The Raven?"
Rectangling the Rectangle [7.14]
    Several dissection tasks using unique dimensions, etc.
Three Geometric Puzzles [5.11, 13.10, 7.2]
    Find the cross; dissect the trapezoid into four equal parts.

December 1979 — Last, Chapter 21

A Poker Puzzle [2.11]
    Straight flush seems easier that four-of-a-kind; find the flaw.
Indian Chess Mystery [12.10]
    Given red-green chess position determine who moved first.
Redistribution in Oilaria [9.24]
    How adjacent classes would redistribute their wealth
Fifty Miles and Hour [4.21]
    If trip averaged 50 mph then some 50 miles in one hour.
A Counter-Jump "Aha!" [10.12]
    Can checkers be moved across a given rectangular board?
Toroidal Paradox [8.18]
    One-hole surface is deformed with a ring.

April 1981 — LAST, Chapter 16

Lavinia Seeks a Room [9.10]
   A set of colinear points; point minimizes sum of distances.
Mirror-Symmetric Solids [6.15]
   What can be rotated into mirror-image without symmetry?
The Damaged Patchwork Quilt [7.4]
   Dissect the 9 × 12 quilt with a hole into a 10 × 10 quilt.
Acute and Isosceles Triangles [7.13]
   A square into nine acute triangles; equilateral into five isoceles.
Measuring with Yen [9.14]
   Measure any unit distance with coins of unit radius.
A New Map-Coloring Game [11.15]
   Find a map which a player can force six colors to be used.
Whim [11.9]
   Nim variant; one move can decide if it is misere or not.

Two additional columns had list of puzzles and these were also collected in the *Colossal Book*. The February 1966 column (Chapter 2 in CARNIVAL) had a list of coin puzzles [problems 5.20, 8.12, 10.3, 10.5, 10.21, 16.4, 16.5]. The July 1969 column (Chapter 2 in CIRCUS) had a list of matchstick puzzles [problems 3.7, 5.3, 5.7, 8.2, 15.3, 16.2, 16.3].

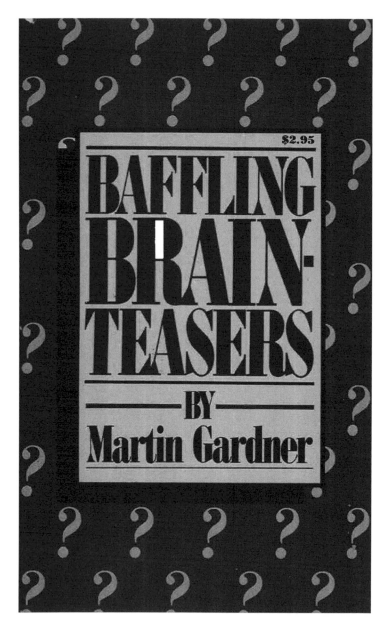

Figure 3.1: Davis Publications, 1983

Chapter 4

The Isaac Asimov Magazine Column

For more than 10 years Gardner contributed a puzzle to *Isaac Asimov's Science Fiction Magazine* (IASFM). The "puzzle" was written in the form a short science fiction scenario and the answer was used as filler at the end of some short story; the answer led to another question whose answer was used as filler for some other story; this continued until three to five fillers were provided. This multi-page puzzle/story approach only would work if Gardner could find three to five related puzzles to include in one scenario. Hence it is remarkable that it appeared in each issue, from the first issue until vol. 10, no. 11, November 1986. The magazine was published irregularly as follows:

- Volume 1 was quarterly; Spring to Winter 1977, nos 1 to 4.
- Volume 2 was bimonthly; Jan/Feb to Nov/Dec 1978, nos 1 to 6.
- Volumes 3 and 4 monthly; Jan to Dec 1979 to 1980, nos 1 to 12.
- Volumes 5 to 10 appear once every four weeks; there were two Aug issues in 1981 and a Mid-Dec issue thereafter, nos 1 to 13.

4.1 Book Collections

Those puzzle stories have appeared in three books. which are detailed below. Only the first page is given; the remaining pages are scattered throughout each issue, as indicated above.

Science-Fiction Puzzle Tales (Potter, 1981). Foreword Isaac Asimov.

1	Lost on Capra	Summer	1977	34
2	The Doctor's Dilemma	Spring	1977	39
3	Space Pool	Fall	1977	34
4	Machismo on Byronia	Winter	1977	48
5	The Case of the Defective Doyles	Jan-Feb	1978	45
6	The Third Dr. Moreau	Mar-Apr	1978	42
7	The Voyage of the Bagel	May-Jun	1978	28
8	The Great Ring of Neptune	Jul-Aug	1978	26
9	The Toroids of Dr. Klonefake	Sept-Oct	1978	29
10	The Postage Stamps of Philo Tate	Nov-Dec	1978	40
11	Captain Tittlebaum's Test	Feb	1979	40
12	Exploring Carter's Crater	Jan	1979	55
13	Pink, Blue, and Green	Mar	1979	47
14	The Three Robots of Professor Tinker	Apr	1979	40
15	How Bagson Bagged a Board Game	May	1979	42
16	The Shop on Bedford Street	Jun	1979	40
17	Tanya Tackles Topology	Jul	1979	60
18	The Explosion of Blabbage's Oracle	Aug	1979	43
19	Dracula Makes a Martini	Sept	1979	42
20	The Erasing of Philbert the Fudger	Nov	1979	46
21	Oulipo Wordplay†	Oct	1979	47
22	How Crock and Witson Cracked a Code‡	Dec	1979	36
23	Titan's Titanic Symbol	Jan	1980	44
24	Professor Cracker's Antitelephone	Feb	1980	36
25	Vacation on the Moon	Mar	1980	38
26	Weird Numbers from Titan	May	1980	42
27	Lucifer at Las Vegas	Apr	1980	48
28	Off We're Going to Shuttle	Jun	1980	45
29	The Backward Banana	Jul	1980	54
30	The Queer Story of Gardner's Magazine	Aug	1980	36
31	Blabbage's Decision Paradox††	Nov	1980	41
32	No Vacancy at the Aleph Null Inn	Sept	1980	28
33	Tube Through the Earth	Dec	1980	40
34	Robots of Oz	Oct	1980	53
35	The Dance of the Jolly Green Digits	Feb	1981	50
36	The Bagel Heads Home	Jan	1981	58

† IASFM title "On Oulipo Algorithms, Anagrams, and Other Nonsense."
‡ IASFM title "How Crock and Watkins Cracked a Code."
†† IASFM title "G. Hovah's Decision Paradox"

Puzzles from Other Worlds (Vintage, 1984).

1	Chess by Ray and Smull	Mar	1981	35
2	The Polybugs of Titan	Apr	1981	45
3	Cracker's Parallel World	May	1981	38
4	The Jinn from Hyperspace	Jul	1981	35
5	Titan's Loch Meth Monster	Jun	1981	45
6	The Balls of Aleph-Null Inn	Aug 3	1981	63
7	Scrambled Heads on Langwidere	Aug 31	1981	47
8	Antimagic at the Number Wall	Sept	1981	57
9	Parallel Pasts	Oct	1981	30
10	Luke Warm at Forty Below	Nov	1981	66
11	The Gongs of Ganymede	Dec	1981	38
12	Tanya Hits and Misses	Jan	1982	37
13	Mystery Tiles at Murray Hill	Feb	1982	40
14	Crossing Numbers on Phoebe	Mar	1982	42
15	SFs and Fs on Fifty-fifth St.†	Apr	1982	56
16	Humpty Falls Again	May	1982	36
17	Palindromes & Primes	Jun	1982	54
18	Thirty Days Hath Sept	Dec	1982	26
19	Home Sweet Home	Jul	1982	24
20	Fingers and Colors on Cromo	Aug	1982	27
21	Valley of the Apes	Sept	1982	27
22	Dr. Moreau's Momeaters	Oct	1982	35
23	And He Built Another Crooked House	Nov	1982	43
24	Piggy's Glasses and the Moon	Mid-Dec	1982	16
25	Monorails on Mars	Jan	1983	28
26	The Demon and the Pentagram	Feb	1983	26
27	Flarp Flips a Fiver	Mar	1983	23
28	Bouncing Superballs	Apr	1983	50
29	Run, Robot, Run	May	1983	22
30	Thang, Thung, and Metagame	Jun	1983	24
31	The Number of the Beast	Jul	1983	22
32	The Jock Who Wanted to Be Fifty	Aug	1983	25
33	Fibonacci Bamboo	Sept	1983	21
34	Tethered Purple-Pebble Eaters	Oct	1983	20
35	The Dybbuk and the Hexagram	Nov	1983	26
36	1984	Dec	1983	27
37	The Castrati of Womensa	Mid-Dec	1983	22

† IASFM title "SFs and Fs on 52nd St."

Riddles of the Sphinx: and Other Mathematical Puzzle Tales (Mathematical Association of America, 1987). (New Mathematical Library, vol. 32)

1	Riddles of the Sphinx†	Jan	1984	28
2	Precognition and the Mystic 7	Feb	1984	26
3	On to Charmian	Mar	1984	20
4	Technology on Vzigs	May	1984	22
5	The Valley of Lost Things	Jun	1984	28
6	Around the Solar System	Aug	1984	20
7	The Stripe on Barberpolia	Sept	1984	22
8	The Road to Mandalay	Jul	1984	28
9	The Black Hole of Cal Cutter	Oct	1984	22
10	Science Fantasy Quiz	Nov	1984	42
11	The Barbers of Barberpolia	Dec	1984	28
12	It's All Done with Mirrors	Mid-Dec	1984	20
13	Satan and the Apple	Jan	1985	52
14	How's-that-again Flanagan?	Feb	1985	26
15	Relativistically Speaking	Mar	1985	20
16	Bar Bets on the Bagel	Apr	1985	21
17	Catch the BEM	May	1985	18
18	Animal TTT	Jun	1985	24
19	Playing Safe on the Bagel	Jul	1985	38
20	Sex Among the Polyomans	Sept	1985	36
21	Inner Planets Quiz	Aug	1985	20
22	Puzzles in Flatland	Oct	1985	20
23	Dirac's Scissors	Nov	1985	20
24	Bull's-Eyes and Pratfalls	Dec	1985	24
25	Flarp Flips Another Fiver	Mid-Dec	1985	28
26	Blues in the Night	Jan	1986	18
27	Again, How's That Again?	Feb	1986	21
28	Alice In Beeland	Mar	1986	16
29	Hustle Off To Buffalo	Apr	1986	22
30	Ray Palmer's Arcade	May	1986	16
31	Puzzle Flags on Mars	Jun	1986	18
32	The Vanishing Plank	Jul	1986	62
33	987654321	Aug	1986	18
34	Time-Reversed Worlds‡	Sept	1986	20
35	The Wisdom of Solomon	Oct	1986	18
36	Thang, the Planet Eater	Nov	1986	17

† IASFM title "Riddles of the Sphinxes."

‡Byline was "Rendrag Nitram."

The puzzle tale "Two Odd Couples" (IASFM, April 1984, pp. 23+) appears to have not be collected. Since they were collected in approximately the same order as their magazine appearance, no inverted listing is given.

4.2   Other Books Containing Puzzles Tales

1. "Machismo on Byronia," *Asimov's Choice: Black Holes & Bug-Eyed-Monsters.* George Scithers (Ed) (Davis Publ, 1977).
2. "The Voyage of the Bagel," *Asimov's Choice: Dark Stars & Dragons.* George Scithers (Ed) (Davis Publ, 1978).
3. "The Great Ring on Neptune," *Asimov's Choice: Extraterrestrials & Eclipses.* George Scithers (Ed) (Davis Publ, 1978).
4. "The Third Dr. Moreau," *Asimov's Choice: Comets & Computers.* George Scithers (Ed) (Davis Publ, 1978).
5. "The Case of the Defective Doyles," *Isaac Asimov's Masters of Science Fiction.* George Scithers (Ed) (Davis Publ, 1978) 45+.
   Paperback edition was called *Isaac Asimov's Science Fiction Anthology 1.*
6. "The Three Robots of Professor Tinker," *Isaac Asimov's Marvels of Science Fiction.* George Scithers (Ed) (Davis Publ, 1979) 56+.
   Paperback edition titled *Isaac Asimov's Science Fiction Anthology 2.*
7. "The Toroids of Dr. Klonefake," *Isaac Asimov's Worlds of Science Fiction.* George Scithers (Ed) (Davis Publ, 1980) 55+.
   Paperback edition titled *Isaac Asimov's Science Fiction Anthology 4.*
8. "Pink, Blue, and Green," *Isaac Asimov's Near Futures and Far.* George Scithers (Ed) (Davis Publ, 1982) 51+.
   Paperback edition titled *Isaac Asimov's Science Fiction Anthology 5.*
9. *Baffling Brainteasers* (Davis Publications, 1983).
   This small book reprints 13 puzzle-tales from *Puzzles from Other Worlds*; in particular, chapters 19, 17, 16, 15, 3, 7, 6, 9, 2, 13, 11, 14, and 21 respectively. Perhaps only available to IASFM subscribers.

10. *The Jinn from Hyperspace* (Prometheus, 2008).

    This anthology reprints four puzzle-tales in chapters 10, 11, 12, and 13 that appeared in July 1981, January 1985, November 1980, and February 1980 respectively, with new postscripts.

11. *When You Were a Tadpole and I Was a Fish* (Hill and Wang, 2009).

    This anthology reprints four puzzle-tales in chapters 3, 9, 13, and 14 that appeared in December 1985, September 1979, August 1979, and November 1979 respectively, with new postscripts.

4.3   Published Correspondence

1. *Isaac Asimov's Science Fiction Magazine* (November 1984) 14.

    Discusses Patricia Moore's response/answers to the May 1984 column.

Part II

Science and Philosophy

Martin Gardner left high school in 1932 expecting to major in physics. Tulsa Central High was extraordinary and prepared him well. The high school had faculty with Ph.D.s and Master's degrees. So he was well-aware that physics was increasingly exciting with new discoveries weekly and that it was based on mathematical principles.

However Caltech did not accept freshman at the time and so he went to the University of Chicago. Chicago had recruited in Oklahoma for the best students. When he arrived at Chicago he was a Protestant fundamentalist. He believed in a young earth as put forward by the creationist George McCready Price. When he took a freshman Geology course he found the evidence against Price too pervasive and persuasive. This changed the direction of his studies. Even though the University of Chicago was nominally a Baptist university he found it hard to hold on to his beliefs. It was a slow dissolution due to coursework and outside reading in both science and philosophy.

He fell under the influence of the Philosophy professors and majored in the Philosophy of Science. Unlike most students (and most faculty) he pursued this direction to understand what he should believe and why. Therefore his efforts were concerted because of the importance to him. He did not believe anything because a professor said it, rather he would seek the underpinnings. (He wrote an editorial in the campus newspaper about this.)

It is curious that while philosophy dominated his intellectual life for the next 75 years he rarely wrote about it, except in his confessional magnum opus *The Whys of a Philosophical Scrivener*. Instead it was evidenced by his critical evaluation of science, first as a philosopher of science. In time he examined fringe science. He found it deeply troubling because his own belief system was bound up with the integrity of science. Further, he had a deep sense of awe and wonder in the natural world and pseudoscience was like graffiti. While he felt it deserved ridicule he found that just describing it would lead his readers to find it ridiculous.

While at Chicago around 1940-1941 he worked with the public relations office and was in charge of press releases about science. This is the roots of his journalistic approach to science. When it became clear that his future was to be a free-lance journalist it is natural that his non-fiction would be in this direction. For example, he wrote an article for *Scientific American* about logic machines in 1952; this was expanded into a book in 1958. His *Relativity for the Millions* and *The Ambidextrous Universe* continued his early fascination with physics. It would never have occurred to him to write a book about botany or chemistry.

Gardner's writings on pseudoscience (or fringe science) started in 1950 with an article called "The Hermit Scientist." The topic was natural due to his personal interest in the integrity of science. He wanted to write a book on hoaxes, but was convinced that topic had been done before but he could expand his article instead. The resulting *In the Name of Science* appeared in 1952 and modern skepticism was born. Gardner did not pursue the subject right after that but when the "occult revolution" of the early 1970s took off, he helped form CSICOP and contributed a column to their magazine for 20 years. This soon dominated his public life; for example, he was asked to write over 40 book reviews about fringe science.

Later in this Part we find Gardner's contributions to the philosophy of science. The earliest articles were outgrowths of his university life and coursework, often with Rudolph Carnap.

The next chapter focuses on the intersection of philosophy and religious belief that was Gardner's original motivation for academic study. It never left him. His displeasure with Christian charlatans was a recurring theme.

Finally, Gardner's interest in politics is explored. His writings were all in support of democratic socialism.

Chapter **5**

Science

His interest in science combined with his success in publishing nonfiction led naturally to his decision to concentrate on "popular science." He would never write about something he did not understand—not as much as an expert but much more than the reader. For example, in his files you can find a book-length typed binder of his notes for the book on relativity. Only after he had worked out an explanation for himself did he feel he could write for the public. Over time his confidence increased, but he never would assert something he did not understand.

5.1 Books

1. *Logic Machines and Diagrams* (McGraw-Hill, 1958).

   - *Logic Machines, Diagrams and Boolean Algebra* (Dover, 1968).
   - *S:GBB* (1981) 27-51.

     Only chapter 1 of the book is included here.

   - *Logic Machines and Diagrams (Second Edition)* (University of Chicago Press, 1982).

     Foreword by Donald Michie. Chapter 9 was rewritten to address machine intelligence.

- *JFH* (2008) 57-82.

  Only chapter 1 of the book is included here.

2. *Relativity for the Million* (MacMillan, 1962).

  - *Relativity for the Million* (Pocket Books, 1965).
  - *The Relativity Explosion* (Vintage, 1976).

    "A completely revised, updated edition"

  - *Relativity Simply Explained* (Dover, 1997).

    "The Dover edition makes new corrections, restores all the 1962 illustrations in their original color, and adds a new Introduction and a new Postscript by the author."

3. *The Ambidextrous Universe* (Basic, 1964).

  Subtitle on jacket was "Left, Right and the Fall of Parity."

  - *The Ambidextrous Universe: Left, Right and the Fall of Parity* (Mentor Book / New American Library, 1969).

    Revised updated edition.

  - *The Ambidextrous Universe: Mirror Symmetry and Time-Reversed Worlds* (Charles Scribner's Sons, 1979).

    Second revised, updated edition.

  - *The New Ambidextrous Universe: Symmetry and Asymmetry from Mirror Reflections to Superstrings* (W. H. Freeman, 1990).

    Third revised edition. "The book is replete with changes throughout, and Chapters 26, 27, 32, 33, and 34 are entirely new." (Three printings differ by the inclusion of corrections.)

  - *The New Ambidextrous Universe: Symmetry and Asymmetry from Mirror Reflections to Superstrings* (Dover, 2005).

    This is revision of the Freeman edition but is still subtitled "Third Revised Edition." Contains a new preface, corrections, and 8 pages of new Notes.

  - *JFH* (2008) 45-55.

    This reprints one chapter, "Time-Reversed Worlds."

5.2 Books Edited

1. [Various Authors], *Great Essays in Science* (Pocket Books, 1957).

   - [Various Authors], *Great Essays in Science* (Washington Square Press, 1961).

2. [Various Authors], *"The Sacred Beetle" and Other Great Essays in Science* (Prometheus, 1984).

   Updated version of the 1957 book, with four new chapters and one chapter removed.

   - *"The Sacred Beetle" and Other Great Essays in Science* (Meridian, 1986).
   - *Great Essays in Science* (Prometheus, 1994).

     Title shortened to the original length.

   - *Great Essays in Science* (Oxford University Press, 1997).

     Reissued "with corrections."

5.3 Book Introductions

1. Roger Penrose, *The Emperor's New Mind* (Oxford University Press, 1989) v-vii.

   - "Reflections on the Emperor's New Mind." *Noetic Sciences Review* 16 (Autumn 1990) 14.
   - *JFH* (2008) 39-43.

5.4 Articles

1. "Logic Machines." *Scientific American* **186** (March 1952) 68-73.
2. "Thought for Today." *The* [Baltimore] *Sun* (March 29, 1952) 8.

   Excerpt from *Scientific American*

3. "The Robots are Coming." *Fate* **5**, 5 (July/August 1952) 34-40.
4. "The World's Greatest Scientific Needs." *P.B.: Pocket Book Magazine* 2 (December 1954) 86-103.
5. "Nature's Mysterious Left-Handed Twist." *Best Articles & Stories* 6 (January, 1960) 52-53.
6. "O Why of the Eye." *Saturday Review* **47** (November 7, 1964) 67.

   Excerpted from *Ambidextrous Universe*

7. "Logic Diagrams." *Encyclopedia of Philosophy* **5** (1967) 77-81.

   - "Logic Diagrams." *Encyclopedia of Philosophy: Second edition* **5** (2006) 560-564.

8. "Logic Machines." *Encyclopedia of Philosophy* **5** (1967) 81-83.

   - "Logic Machines." *Encyclopedia of Philosophy: Second edition* **5** (2006) 565-567.

9. "The Message from Planet X." In *Philosophy*, edited by James L. Christian (Holt, Rinehart and Winston, 977) 540.

   Excerpted from *Relativity for the Million*; further the excerpt "And Then What Happened?" appears on page 570.

10. "Great Fakes of Science." *Esquire* (October 1977) 88-91.

    - *S:GBB* (1981) 123-133.
    - "Great Fakes of Science." In *The Faber Book of Science*, edited by John Carey (Faber and Faber, 1995) 451-454.
    - "Great Fakes of Science." In *Eyewitness to Science*, edited by John Carey (Harvard University Press, 1995) 451-454.

11. "The Computer as Scientist." *Discover* **4**, 7 (June 1983) 85-88.

    - *W&W* (1989) 48-56.
    - *TNIL* (1996) 32-39.

12. "Illusions of the Third Dimension." *Psychology Today* (August 1983) 62-67.

    - *W&W* (1989) 57-67.

13. "Guest Comment: Is Realism a Dirty Word?" *American Journal of Physics* **57**, 3 (March 1989) 203.

    - "Thinking about the Mind and the Universe." *Skeptical Inquirer* (Summer 1990) 446. Excerpt.
    - *OTWS* (1992) 182-184.

14. "A Tribute [to Isaac Asimov]." *Skeptic* **1**, 1 (Spring 1992) .

15. "A Celebration of Isaac Asimov: A Man for the Universe." *Skeptical Inquirer* **17**, 1 (Fall 1992) 30-45. (Gardner's contribution is on page 42.)

16. "Computers Near the Threshold?" In *Mysteries of Life and the Universe: New Essays from America's Finest Writers on Science*, edited by William H. Shore (Harcourt Brace Jovanovich, 1992) 126-135.

    - "Computers Near the Threshold?" *Journal of Consciousness Studies* **3**, 1 (1996) 89-94. Slightly updated.

- *TNIL* (1996) 442-449.
- *GW* (2001) 97-107.

17. "The Darkened Cosmos: A Tribute to Carl Sagan." *Skeptical Inquirer* **21**, 2 (March/April 1997) 5-15. (Gardner's contribution is on page 7.)

18. "The Fourth Dimension: An Excerpt from *The Ambidextrous Universe.*" In *Metaphysics: The Big Questions*, edited by Van Inwagen, Peter and Dean W. Zimmerman (Blackwell Publ, 1998) 108-111.

5.5   Book Reviews

1. "What's Up, Doc?" *Saturday Review* (April 18, 1953) 15,44.

   Preview for Tomorrow, Bruce Bliven

2. *Book News: Library of Science* **6**, 1 (Sept. 1959) 5-7. Advertising brochure.

   Electronic Data Processing, R. Nett and S. Hetzler

3. *Library of Science* (ca. 1973). Advertising brochure.

   Random Walk in Science, Edited by R. L. Weber

4. "The Tiny Computer Inside Our Skull." *Washington Post Book World* (August 2, 1970) 4.

   The Intelligent Eye, R.L. Gregory

   - *O&S* (1983) 269-271.

5. "Monkey See, Monkey Do, Monkey Say: The Language of Learning." *Washington Post* (March 18, 1975) B5.

   Apes, Men and Language, Eugene Lindon

6. "The Man Who Invented the H-Bomb." *New York Times Book Review* (May 9, 1976) 8.

   Adventures of a Mathematician, S.M. Ulam

   - *O&S* (1983) 298-304.

7. "Bang!" *New York Review of Books* (May 12, 1977) 29-31.

   The First Three Minutes, Steven Weinberg

   - *O&S* (1983) 313-319.

8. "The Holes in Black Holes." *New York Review of Books* (September 29, 1977) 22-24.

   The Collapsing Universe, Isaac Asimov
   Space, Time and Gravity, R.M. Wald
   The Key to the Universe, Nigel Calder
   Ten Faces of the Universe, Fred Hoyle
   The Iron Sun, Adrian Berry
   White Holes, John Gribbin
   Space and Time in the Modern Universe, P.C.W. Daires

   • *S:GBB* (1981) 335-346.

9. "O Brave New World." *New York Times Book Review* (October 23, 1977) 11, 42.

   The People Shapers, Vance Packard

   • *O&S* (1983) 323-325.

10. "The Charms of Catastrophe." *New York Review of Books* (June 15, 1978) 30-33.

    Structural Stability and Morphogenesis, Rene Thom
    Catastrophe Theory, E.C. Luman
    Catastrophe Theory, A. Woodcock and M. Davis
    Catastrophe Theory and Its Application, T. Potson and I. Stewart

    • *S:GBB* (1981) 365-374.

11. "Hello and Goodbye." *New York Review of Books* (November 23, 1978) 12-16.

    Murmurs of the Earth, Carl Sagan et al.
    Spaceships of the Mind, Nigel Calder
    In the Center of Immensities, B. Lovell

    • *O&S* (1983) 328-335.

12. "Keeping Up with Einstein." *New Leader* (May 21, 1979) 22-24.

    Einstein, L. Broglie et al

    • *O&S* (1983) 336-338.

13. "Eternal Riddles." *New York Review of Books* (June 14, 1979) 32-34.

    Broca's Brain, Carl Sagan

    • *S:GBB* (1981) 379-390.

14. "Monkey Business." *New York Review of Books* (March 20, 1980) 3-5.

Nim, Herbert Terrace
Speaking of Apes, edited by T.A. Sebeok

- *S:GBB* (1981) 391-408.

15. "The World of Quantum." *Science Digest* (June 1981) 106.

Other Worlds, Paul Davies

16. "Bottled Mischief." *Science Digest* (September 1981) 124.

Polywater, Felix Franks

17. "The Wonders of Old China." *Science Digest* (March 1982) 37.

Science in Traditional China, Joseph Needham

- *W&W* (1989) 157-158.

18. "How Well Can Animals Converse?" *Semiotica* **38**, 3/4 (1982) 357-367.

The Clever Hans Phenomenon, T.A. Sebeok and R. Rosenthal

- *O&S* (1983) 380-390.

19. "A Trail of Genius." *Science 83* (September 1983) 96-97.

Great Scientific Experiments, Rom Harre

- *W&W* (1989) 159-161.

20. "A Feeling of the Unfathomable." *Nature* **308** (April 26, 1984) 785-786.

Reality and Empathy, Alex Comfort

- *W&W* (1989) 175-178.

21. "The Man Who Invented the Future." *Science Digest* (June 1984) 80-81.

Ascent to Orbit: A Scientific Autobiography, Arthur C. Clarke

- *W&W* (1989) 188-190.

22. "Life with a Lighter Touch." *Nature* **314** (April 25, 1985) 685-686.

    Surely You're Joking, Mr. Feynman, Richard Feynman with Ralph Leighton

    - *W&W* (1989) 194-196.

23. "Physics: The End of the Road?" *New York Review of Books* (June 13, 1985) 31-34.

    Superforce, Paul Davies
    Perfect Symmetry, Heinz Pagels

    - *W&W* (1989) 197-210.

24. "Secrets of the Old One." *New York Review of Books* (December 4, 1986) 49-50.

    Was Einstein Right?, Clifford M. Will

    - *W&W* (1989) 229-237.

25. "Cataloging the Mind's Hardware." *Boston Sunday Globe* (February 8, 1987) A11, A13.

    The Society of Mind, Marvin Minsky

    - *W&W* (1989) 238-242.

26. "Finding Order in Chaos in Our Physical Worlds." *Boston Sunday Globe* (October 4, 1987) C36-C37.

    Chaos, James Gleick

    - *W&W* (1989) 243-246.

27. *News and Observer* [Raleigh, NC] (April 10, 1988) 4D.

    Infinite in All Directions, Freeman Dyson

    - *WW&FL* (1996) 175-177.

28. "And if the Robot Fouls Up, Well, That's Life." *News and Observer* [Raleigh, NC] (October 16, 1988) 4D.

    Mind Children, Hans Moravec

    - *OTWS* (1992) 153-154.

29. "The Ultimate Turtle." *New York Review of Books* (June 16, 1988) 17-20.

   A Brief History of Time, Stephen W. Hawking

   - *TNIL* (1996) 92-102.

30. *News and Observer* [Raleigh, NC] (June 19, 1988).

   Interactions, Sheldon L. Glashow

   - *WW&FL* (1996) 178-180.

31. "Playing Computer Games with Reality." *News and Observer* [Raleigh, NC] (July 17,1988).

   The Dreams of Reason, Heinz Pagels

   - *WW&FL* (1996) 181-183.

32. "Physics from A to Zee." *News and Observer* [Raleigh, NC] (February 26, 1989) 5D.

   Fearful Symmetry, Anthony Zee

   - *WW&FL* (1996) 184-185.

33. "An Inventive American Century." *News and Observer* [Raleigh, NC] (April 16, 1989) 5D.

   American Genesis, Thomas Hughes

   - *WW&FL* (1996) 193-194.

34. "To Pluto and Beyond: The New Frontier." *News and Observer* [Raleigh, NC] (July 23, 1989).

   Journey into Space, Bruce Murray

   - *WW&FL* (1996) 195-197.

35. "To Follow Knowledge Like a Sinking Star." *Washington Post Book World* (September 2, 1990) 4-5.

   Origins, Alan Lightman and Roberta Brawer

   - *WW&FL* (1996) 198-200.

36. "Written in the Stars." *Washington Post Book World* (January 20, 1991) 5.

   Lonely Hearts of the Cosmos, Dennis Overbye

   - *WW&FL* (1996) 209-211.

37. "Those Little Gray Cells." *Washington Post Book World* (February 17, 1991) 1-2.

    In the Palaces of Memory, George Johnson

    - *WW&FL* (1996) 212-215.

38. "The Myriad Worlds of a Naturalist." *Washington Post Book World* (May 19, 1991) 1,10.

    Bully for Brontosaurus, Stephen Jay Gould

    - *WW&FL* (1996) 219-221.

39. "Feynman the Mad Magician." *News and Observer* [Raleigh, NC] (October 25, 1992) 4G.

    Genius, James Gleick

    - *WW&FL* (1996) 245-247.

40. "Inside Looking Out." *Washington Post Book World* (March 22, 1992) 5.

    The Mind's Sky, Timothy Ferris

    - *WW&FL* (1996) 237-239.

41. "Freeman Dyson and Human Adventure." *Washington Post Book World* (July 5, 1992) 4.

    From Eros to Gaia, Freeman Dyson

    - *WW&FL* (1996) 242-244.

42. "The Heart of the Matter." *Washington Post Book World* (January 31, 1993) 9.

    Dreams of a Final Theory, Steven Weinberg

    - *WW&FL* (1996) 255-257.

43. "Informed Opinions: Experts Choose Their Favorite Books." *Washington Post Book World* (December 4, 1994) 12.

    Capsule reviews of 6 books.

44. "Kilroy was Here." *Los Angeles Times Book Review* (March 5, 2000) 4-5.

    The Meme Machine, Susan Blackmore

    - *WJ* (2000) 207-216.

45. "Theory of Everything." *New Criterion* (October 2004) 65-68.

The Road to Reality, Roger Penrose

- *JFH* (2008) 33-38.

46. "Science in the Looking Glass: What Do Scientists Really Know?" *Notices of the AMS* **52**, 11 (December 2005) 1344-1347.

Science in the Looking Glass, E. Brian Davies

- *JFH* (2008) 93-103.

47. "M Is for Messy." *New Criterion* (April 2007) 65-68.

The Trouble with Physics, Lee Smoil
Not Even Wrong, Peter Woit

- *JFH* (2008) 141-146.

48. "Is Beauty Truth and Truth Beauty." *Scientific American* (April 2007) 84-85.

Why Beauty is Truth: A History of Symmetry, Ian Stewart

- *JFH* (2008) 137-140.

49. "Do Loops Explain Consciousness?" *Notices of the AMS* **54**, 7 (August 2007) 84-85.

I Am a Strange Loop, Douglas Hofstadter

- *JFH* (2008) 147-152.

50. "Sir Isaac's Ocean." *New Criterion* (April 2008) 75-77.

Newton, Peter Ackroyd

- *T&F* (2009) 9-13.

5.6   Published Correspondence

1. "Inventions." *New York Review of Books* (July 10, 1969) 36.

Discusses *Ambidextrous Universe* and Nabokov.

2. "Only Joking." *New York Review of Books* (Dec. 8, 1977) 54.

A reply to John Gribbin concerning black holes.

- *S:GBB* (1981) 344-345.

3. *Physics Today* (November 1985) 136.

Proposes a puzzle version of the EPR paradox

4. "The Michelson-Morley Experiment." *New York Review of Books* (May 12, 1988) 61.

Rejoinder about a remark about the Michelson-Morley experiment.

5. "The Meaning of Quantum Theory." *Physics and Society* **18**, 4 (October 1989) 4.

A reply to an editorial by Art Hobson criticizing Gardner's comments in the *American Journal of Physics*, followed by a rejoinder.

- *OTWS* (1992) 184-185.
- *TNIL* (1996) 424.

6. "The Meaning of Quantum Theory, Cont." *Physics and Society* **19**, 1 (October. 1989) 2

A reply to Art Hobson about the lack of "fresh metaphysical questions" in quantum mechanics.

- *OTWS* (1992) 185-187.
- *TNIL* (1996) 424-426.

7. "Body and Soul." *Washington Post Book World* (April 14, 1991) 14.

Rejoinder to George Johnson, the author of *Palaces of Memory*.

- *WW&FL* (1996) 214-215.

8. *The Physics Teacher* **31** (September 1993) 330.

Comments on the futility of avoiding the phrase "wave function."

9. *Analog Science Fiction & Fact* **114**, 12 (October 1994) 174.

Corrects assertions by John Cramer about the planet Mercury.

10. *Commentary* **102**, 3 (September 1996) 15-16.

Replies to David Berlinski's attack on evolution.

11. *Skeptical Inquirer* (September/October 2000) 67.

Rejoinder to letters about Bohm's guided wave theory.

12. "By the Book." *Science News* **172**, 9 (September 1, 2007) 143.

Correction to an assertion about Guth's inflation model.

Chapter **6**

Fringe Science

He wrote a lot about science, but first made his mark by focusing on pseudo-science, a topic he had kept up with since college. A pivotal article was "The Hermit Scientist" (1950). In that article he critiqued three examples of pseudo-science that were very much in the news: Velikovsky, the UFO craze, and Dianetics. This led to the book *In the Name of Science* (1952). In 1972 Gardner, James Randi and Ray Hyman decided that a group was needed to address what was then called the "occult explosion," an alarming escalation of irrationalism in the popular press. Soon they were joined by Marcello Truzzi and Paul Kurtz and the Committee for the Scientific Investigation of Claims of the Paranormal (CSICOP) was formed. He contributed to their journal regularly for decades; there is a separate chapter for the column he wrote for them.

Gardner is often called the "founder of the modern skeptical movement." It is not because he was a great organizer. It is because his clearly-written articles drew people in, they convinced us there was a problem we should care about, they showed us how simple argumentation was effective in combatting unreason, and they gave a common voice to those who were reluctant to speak out.

His antipathy for pseudo-scientists was not just because they were cranks, crackpots or charlatans. He was furious about those who practiced medical hocus-pocus. In my opinion, his motivation was almost aesthetic. He was offended, deeply offended, by people offering "ugly" false phenomena and miracles when the natural world was so wonderful.

6.1   Books

To avoid confusion, books of collected columns from *The Skeptical Inquirer* are listed here even though they are discussed more fully in a later chapter.

1. *In the Name of Science* (Putnam, 1952).

   - *Fads and Fallacies in the Name of Science* (Dover, 1957).
     Revised edition, with new preface.
   - *Fads and Fallacies in the Name of Science* (Ballantine, 1957).
     Abridged edition (16 out of 26 chapters) based on the Dover edition.
   - *Fads and Fallacies in the Name of Science* (New American Library, 1986).
     Hardcover reprint in the "Masterpieces of Science" series

2. Uriah Fuller [Martin Gardner], *Confessions of a Psychic* (Karl Fulves, 1975).

   A pamphlet for magicians; also listed in the Magic part.

3. Uriah Fuller [Martin Gardner], *Further Confessions of a Psychic* (Karl Fulves, 1980).

   A pamphlet for magicians; also listed in the Magic part.

4. *Science: Good, Bad, and Bogus* (Prometheus, 1981).

   Contains virtually every article and book review on pseudo-science and parapsychology up to 1981. Note the chapter "Great Fakes of Science" is the original longer version of the *Esquire* article.

   - *Science: Good, Bad, and Bogus* (Discus/Avon, 1983).
   - *Science: Good, Bad, and Bogus* (Prometheus, 1996).
     Audio tape reading by Eric Burns; 180 minutes included a short introduction read by Gardner

5. *The New Age: Notes of a Fringe-Watcher* (Prometheus, 1988).

   Contains 19 of the *Skeptical Inquirer* columns and 14 reviews and essays.

6. *How Not to Test a Psychic: Ten Years of Remarkable Experiments with Pavel Stepanek* (Prometheus, 1989).

7. *On the Wild Side* (Prometheus, 1992).

   Contains 16 of the *Skeptical Inquirer* columns and 15 reviews and essays.

8. *Weird Water & Fuzzy Logic: More Notes of a Fringe Watcher* (Prometheus, 1996).

   Contains 16 of the *Skeptical Inquirer* columns and 30 reviews.

9. *Did Adam and Eve Have Navels?* (Norton, 2000).

   Contains 27 of the *Skeptical Inquirer* columns plus one article.

10. *Are Universes Thicker than Blackberries?* (Norton, 2003).

    Contains the final 12 *Skeptical Inquirer* columns plus many other articles on science, mathematics, science and religion, some previously unpublished.

11. *Dear Martin / Dear Marcello: Gardner and Truzzi on Skepticism* (World Scientific, 2017).

    This is the unedited correspondence between Martin Gardner and Marcello Truzzi over many years. It centers on the "demarcation problem," the dividing line between good science and bad science. Edited by Dana Richards with a lengthy introduction.

6.2 Books Edited

1. *The Wreck of the Titanic Foretold?* (Prometheus, 1986).

   Includes a long introductory essay on such literary coincidences and contains three stories and some poems.

   - *The Wreck of the Titanic Foretold?* (Prometheus, 1998).
     Contains a new preface and "several minor errors" corrected.
   - *T&F* (2009) 75-99.
     Reprints the Introduction

2. H. G. Wells, Hilaire Belloc, *Mr. Belloc Objects and Still Objects* (Battered Silicon Dispatch Box, 2008).

   Contains a foreword, two introductions, and an epilogue

6.3   Book Introductions

1. David Marks and Richard Kamman, *The Psychology of the Psychic* (Prometheus, 1980) 1-3.
2. Ronald Ecker, *Dictionary of Science and Creationism* (Prometheus, 1990) v-viii.
3. Gordon Stein, *Encyclopedia of Hoaxes* (Gale Research, 1993) xiii-xiv.
4. Theodore Schick, Jr, Lewis Vaughn, *How to Think about Weird Things* (Mayfield, 1995) v-vi. (also Second Edition, 1995)

   - Theodore Schick, Jr, Lewis Vaughn, *How to Think about Weird Things: Third Edition* (McGraw-Hill Higher Education, 2002) v-vi. (also Fourth Edition, 2005, and Fifth Edition, 2008)

6.4   Columns

1. "Notes of a Psi-Watcher." *Skeptical Inquirer* **7**, 4 (Summer 1983).

For more details see next chapter.

6.5   Articles

Articles that reprint *Skeptical Inquirer* columns are not listed, they appear in the chapter detailing that column.

1. "It is Smart to be Fooled." *Extensions* **42**, 8 (January 1948) 424-425.
2. "The Hermit Scientist." *The Antioch Review* **10** (1950) 447-457.
   - *S:GBB* (1981) 3-14.
   - "The Hermit Scientist." *The Antioch Review* **74** (Summer 2016) 545-555. 75th Anniversary issue
3. "Phooey on Nostradamus!" *Startling Stories* **27**, 3 (October 1952) 41-42.

This is contained within a novella, "Asylum Earth," by Gardner's friend Bruce Elliott. Gardner could not recall if he wrote it, but thought Elliott put his name on it since it borrows heavily from *In the Name of Science*.

4. "In the Name of Science." *Phoenix* 272 (January 9, 1953) 1089.

   Excerpted from *In the Name of Science*.

5. "That Dowsing Hokum." *Science Digest* **33** (March 1953) 30-36.

   Excerpted from *In the Name of Science*.

   - "Sornettes du Sourcier." *Sciences Selection des Publications Scientifiques* (Mai 1953) 28-34.

6. "The Great Pyramid Hoax." *Science Digest* **33** (May 1953) 10-14.

   Excerpted from *In the Name of Science*.

7. "Quacks, Fads & Cultists." *Everybody's Digest* (May 1953) 82-89.

   Excerpted from *In the Name of Science*.

8. "What Happened to Dianetics?" *Why* **1**, 16 (June 1953) 34-39, 121-122.

   Excerpted from *In the Name of Science*.

9. "Unter Falscher Flagge [Under False Colors]." *Der Monat* 62 (November 1953) 162-171.

   Excerpted from *In the Name of Science*.

10. "The Quest for Perpetual Motion." *Science World* **2**, 1 (September 24, 1957) 10-11.

11. "In the Name of Science." In *Thought and Statement*, edited by W. G. Leary and J. S. Smith (Harcourt, 1960) 204-213.

    Excerpted from *In the Name of Science*.

12. "Anyone You Know?" In *A Dover Science Sampler*, edited by George Barkin (Dover, 1959) 32-33.

    Excerpted from *Fads and Fallacies in the Name of Science*.

13. "In the Name of Science." In *Science: Method and Meaning*, edited by S.B. Rapport (New York University Press, 1963) 31-42.

    Excerpted from *In the Name of Science*.

14. "Manifesto of the Institute of General Eclectics." *Stranger than Fact* [S-F Fanzine] **2**, 1 (Summer 1964) 13-14.

    - "Manifesto of the Institute of General Eclectics." *The Worm Runner's Digest* **3**, 1 (April 1966) 99-100.
    - *S:GBB* (1981) 59-61.

15. "New Trends in Pseudo-Science." *The Quid* [Thomas Jefferson High School, Elizabeth NJ] **1**, 2 (1964) 17-19.

    - *S:GBB* (1981) 53-57.

16. "Charles Fort and the 'Priests of Science'." *Fate* **18**, 8 (August 1965) 90-96.

    Excerpted from *Fads and Fallacies in the Name of Science*.

17. "ESP and PK." In *Controversy*, edited by Daniel McDonald (Chandler, 1966) 229-237.

    Excerpted from *In the Name of Science*.

18. "Dermo-Optical Perception: A Peek Down the Nose." *Science* **151** (February 11, 1966) 654-657.

    - "Dermo-Optical Perception: A Peek Down the Nose." In *Research in Psychology*, edited by B.J. Kintz, et al (Scott Foresman and Co., 1970) 32-39.
    - "Dermo-Optical Perception: A Peek Down the Nose." In *Current Research in Psychology*, edited by H.C. Lindgren, et al (Wiley, 1971) 190-198.
    - "Dermo-Optical Perception: A Peek Down the Nose." In *Psychology in the World Today (Second Edition)*, edited by Robert Guthrie (Addison-Wesley, 1971) 186-194.
    - *S:GBB* (1981) 63-73.

19. "Lysenkoism." In *Contours of Experience*, edited by W. Goldhurst (Prentice-Hall, 1967) 233-244.

    Excerpted from *In the Name of Science*.

20. "A Short but Wild History of Spirit Photography." *Popular Photography* **61** (October 1967) 65.

21. "Scientific Cranks." In *The Project Physics Course: Reader 3: The Triumph of Mechanics* (Holt, Rinehart and Winston, 1971) 248-253.

    Excerpted from *In the Name of Science*.

22. "ESP et PK [in French]." *Les Cahiers Rationalistes* 314 (Janvier 1975) 113-129.

    Excerpted from *Fads and Fallacies*; chapter 25 of the 1966 French translation.

23. "Concerning an effort to demonstrate extrasensory perception by machine." *Scientific American* **233** (October 1975) 114-119.

    Actually a "Mathematical Games" column.

24. "Gardner on Pseudoscientists." In *Persuasion: Understanding, Practice, and Analysis*, edited by Herbert Simon (Addison-Wesley, 1976) 332-333.

Excerpted from *In the Name of Science.*

25. "The Irrelevance of Conan Doyle." In *Beyond Baker Street*, edited by Michael Harrison (Bobbs-Merrill, 1976) 123-135.

- *S:GBB* (1981) 113-122.
- *The Northwest Skeptic* 6 (August 1985) 6-10.
- *TNIL* (1996) 183-192.

26. "Magic and Paraphysics." *Technology Review* **78** (June 1976) 43-51.

- *S:GBB* (1981) 91-112.
- "Magic and Paraphysics." In *Becoming a Successful Student*, edited by Laraine E. Fleming and Judith Leet (Scott, Foresman and Company, 1989).

Excerpted with study questions

27. "Geller, Gulls and Nitinol." *The Humanist* (May/June 1977) 25-32.

- *S:GBB* (1981) 159-178.

28. "A Skeptics View of Parapsychology." *The Humanist* (November/December 1977) 45-46.

- *S:GBB* (1981) 141-150.

29. "Einstein on ESP." *The Zetetic* (Fall/Winter 1977) 53-56.

- *S:GBB* (1981) 151-154.
- "Einstein on ESP." *Indian Skeptic* **3**, 4 (August 1990) 34-36.

30. "The Second Einstein ESP Letter." *The Skeptical Enquirer* (Spring/Summer 1978) 82.

- "Einstein and ESP." In *Paranormal Borderlands of Science*, edited by Kendrick Frazier (Prometheus, 1981) 60-65.

This combined the two previous short articles.

- *S:GBB* (1981) 155-157.
- "The Second Einstein ESP Letter." *Indian Skeptic* **3**, 4 (August 1990) 38-40.

31. "How Does One Distinguish Science from Pseudo-Science." In *Philosophy and Science*, edited by F. E. Mosedale (Prentice-Hall, 1979) 385-391.

Excerpted from *Fads and Fallacies in the Name of Science.*

32. "Quantum Theory and Quack Theory." *New York Review of Books* (May 17, 1979) 39+.

- *S:GBB* (1981) 185-206.
- "Quantum Theory and Quack Theory." *Resonance* (January 2013) 97-101.

  Followed by a piece by J. A. Wheeler

33. "The Extraordinary Mental Bending of Professor Taylor." *The Skeptical Inquirer* (Winter 1979/1980) 67-72.

- "The Extraordinary Mental Bending of Professor Taylor." In *Paranormal Borderlands of Science*, edited by Kendrick Frazier (Prometheus, 1981) 142-147.
- *S:GBB* (1981) 179-184.
- "Selection 204." In *A Literary Companion to Science*, edited by Walter Gratzer (W. W. Norton, 1989) 460-462.

  An excerpt

34. "An Expense of Spirit." *New York Review of Books* (May 15, 1980) 42-43.

- *S:GBB* (1981) 207-214.

35. "Parapsychology and Quantum Mechanics." In *Science and the Paranormal*, edited by George O. Abell and Barry Singer (Scribners, 1981) 56-69.

- "Paraphysicists and Deception." *Skeptical Inquirer* **6**, 4 (Summer 1982) 585-598.

  A short excerpt from the above, used as filler

- "Parapsychology and Quantum Mechanics." In *A Skeptic's Handbook of Parapsychology*, edited by Paul Kurtz (Prometheus, 1985) 585-598.

36. "The Power and the Gory." *Discover* **3**, 8 (August 1982) 8.

Criticizes the use of D. Scott Rogo and other parapsychologists in promoting the Steven Speilberg movie *Poltergeist*.

- *O&S* (1983) 369-371.

37. "How Not to Test a Psychic: The Great SRI Die Mystery." *Skeptical Inquirer* **7**, 2 (Winter 1982/1983) 33-39.

- *NA* (1988) 137-144.
- "How Not to Test a Psychic: The Great SRI Die Mystery." In *Science Confronts the Paranormal*, edited by Kendrick Frazier (Prometheus Books, 1986) 176-181.

- "How Not to Test a Psychic: The Great SRI Die Mystery." *Indian Skeptic* **3**, 4 (August 1990) 11-17.

38. "Great Moments in Pseudo-Science." *Isaac Asimov's Science-Fiction Magazine* **7**, 7 (July 1983) 67-76.

   - *O&S* (1983) 211-216.

39. "Cruel Deception in the Philippines." *Discover* (August 1984) 8.

   - *NA* (1988) 167-169.
   - *WJ* (2000) 32-36.

40. "Perpetual Motion: Illusion and Reality." *Foote Prints* **47**, 2 (1984) 21-35. (Foote Mineral Company.)

   - "Perpetual Motion: The Quest for Machines that Power Themselves." *Science Digest* (October 1985) 68-72,103.
   - *NA* (1988) 145-166.

41. "666 and All That." *Discover* (February 1985) 34-35.

   - *NA* (1988) 170-174.
   - "666." *Magical Blend* 34 (1992) 54-57.
   - "666." In *Encyclopedia of Millennialism and Millennial Movements* , edited by Richard Landes (Routledge, 2000) 1-3.

   Loosely paraphrased, perhaps by the editor.

42. "Science-Fantasy Religious Cults." *Free Inquiry* (Summer 1987) 31-35.

   - *NA* (1988) 209-219.

43. "Evidence of the Paranormal: A Skeptic's Reactions." *Behavioral and Brain Sciences* **10**, 4 (1987) 587-588.

   One of many reactions to position papers by Rao and Palmer, and Alcock.

44. "A View From the Fringe." *Utne Reader* (July/August 1988) 79, 82-83.

   Excerpted from *The New Age*.

45. "Psychic Astronomy." *Free Inquiry* (Winter 1987/1988) 26-33.

   - *NA* (1988) 252-263.

46. "Selection 192 and 194." In *A Literary Companion to Science*, edited by Walter Gratzer (W. W. Norton, 1989) 439-442, 443-446.

   Two excerpts from *Fads and Fallacies*, from the Introduction and Chapter 21, respectively. See also "The Extraordinary Mental Bending of Professor Taylor."

47. "How to Test Your PK Power." *Physics Teacher* (April 1990) 228-229. Continued in the May issue.

48. "PK Power in the Light of May." *Physics Teacher* (May 1990) 296-297.

   - *OTWS* (1992) 203-205. Combines the April and May articles.

49. "How to Fabricate a PPO." *Journal of the Society for Psychical Research* **58**, 824 (July 1991) 43-44.

   - *WW&FL* (1996) 225-227.

50. "Smith and Blackburn: Hornswogglers Extraordinaire." *The Skeptic* (September/October 1991) 12-16.

51. "Communicating with the Dead: William James and Mrs. Piper, Part 1." *Free Inquiry* **12**, 2 (Spring 1992) 20-27.

52. "Communicating with the Dead: William James and Mrs. Piper, Part 2." *Free Inquiry* **12**, 3 (Summer 1992) 38-48.

   - *OTWS* (1992) 217-248.

     Combines parts 1 and 2.

   - *TNIL* (1996) 213-243.

     Combines parts 1 and 2.

53. "Course of Miracles." *The Epistle: The Newsletter of the First Baptist Church of Montclair* [NJ] (May 1993) 2-6.

   Reprint of SI column (Fall 1992) used with permission. Gardner's connection is also shown by his sending in an old poem to the April 1993 *Epistle* p. 3.

54. "Eyeless Vision." In *The Encyclopedia of the Paranormal*, edited by Gordon Stein (Prometheus Books, 1996) 254-264.

   - *UTB* (2003) 225-243.

55. "Magic and Psi." In *The Encyclopedia of the Paranormal*, edited by Gordon Stein (Prometheus Books, 1996) 381-385.

   - *UTB* (2003) 244-251.

56. "*Oahspe.*" In *The Encyclopedia of the Paranormal*, edited by Gordon Stein (Prometheus Books, 1996) 465-470.

   - *UTB* (2003) 101-111.

57. "Mrs. Leonora Piper." In *The Encyclopedia of the Paranormal*, edited by Gordon Stein (Prometheus Books, 1996) 534-539.

   - "How Mrs. Piper Bamboozled William James." In *UTB* (2003) 252-262.

58. " 'Dr.' Henry Slade." In *The Encyclopedia of the Paranormal*, edited by Gordon Stein (Prometheus Books, 1996) 701-706.

- " 'Dr.' Henry Slade, American Medium." In *UTB* (2003) 263-272.

59. "Guest Comment: CSICOP—The Committee for the Scientific Investigation of the Paranormal." *American Journal of Physics* **65**, 4 (April 1997) 273-274.

60. "Life Magazine's Star-Struck View of Astrology." *Skeptical Inquirer* **21**, 6 (November/December 1997) 5-6.

Followed by a response by Jay Lovinger and a rejoinder by Gardner, on p. 8

- *WJ* (2000) 25-28.

61. "Confessions of a Skeptic." In *Skeptical Odysseys*, edited by Paul Kurtz (Prometheus, 2001) 355-362.

62. "Why I Am Not a Paranormalist." In *When You were a Tadpole I was a Fish* (Hill and Wang, 2009) 29-48.

Chapter 3 of *Whys*

63. "The Vagueness of Krishnamurti." *The Ojai Orange* 35 (February 2005) 1, 3-6. ("John Wilcock's personal magazine")

Reprint of *Skeptical Inquirer* column of July 2000.

64. "The Memory Wars: Part 1." *Skeptical Inquirer* **30**, 1 (January/February 2006) 28-31.

- *JFH* (2008) 13-19.

65. "The Memory Wars: Parts 2 and 3." *Skeptical Inquirer* **30**, 2 (March/April 2006) 46-50.

- *JFH* (2008) 21-31.
- "The Memory Wars." In *Science Under Siege*, edited by Kendrick Frazier (Prometheus, 2009) 133-138.

66. "Congratulations from the World." *Free Inquiry* (June/July 2006) 27.

One of many encomiums for "Paul Kurtz at Eighty."

67. "Three Letters." *The Ojai Orange* 51 (Winter 2006). ("John Wilcock's personal magazine")

- *JFH* (2008) 133-136.

68. "'Dr.' Bearden's Vacuum Energy." *Skeptical Inquirer* **31**, 1 (January/February 2007) 18-19.

    - "'Dr.' Bearden's Vacuum Energy." In *Science Under Siege*, edited by Kendrick Frazier (Prometheus, 2009) 311-314.

69. "Arthur C. Clarke Remembered." *Skeptical Inquirer* **32**, 4 (July/August 2008) 30.

70. "Mr. Belloc Objects." *Skeptical Inquirer* **32**, 6 (November/December 2008) 16, 25.

    - H. G. Wells, Hilaire Belloc, *Mr. Belloc Objects and Still Objects* (Battered Silicon Dispatch Box, 2008) 7-9, 11-12, 65-68, 109-112.

      Expanded into a foreword, introductions, and an epilogue

    - *T&F* (2009) 19-26.

      Contains the Introduction ("Objects") and epilogue ("Still Objects").

71. "Why I Am Not a Paranormalist." In *When You were a Tadpole I was a Fish* (Hill and Wang, 2009) 29-48.

    Chapter 3 of *Whys*

72. "Oprah Winfrey: Bright (but Gullible) Billionaire." *Skeptical Inquirer* **34**, 2 (March/April 2010) 54-56.

73. "Lessons of a Landmark PK Hoax." *Skeptical Inquirer* **38**, 6 (November/December 2014) 14-15.

    A reprint of the first column to celebrate Gardner's centennial.

6.6   Book Reviews

1. "Exploding Our Illusions." *Saturday Review* **37** (November 13, 1954) 13.

   The Spoor of Spooks and Other Nonsense, Bergen Evans

   - *O&S* (1983) 249-250.

2. "Funny Coincidence." *New York Review of Books* (May 26, 1966) 27-29.

   ESP: A Scientific Evaluation, C.E.M. Hansel

   - *S:GBB* (1981) 217-231.

3. *Library of Science* (1970). Advertising brochure.

   ESP, Seers, and Psychics, Milbourne Christopher

   - *S:GBB* (1981) 237-239.

4. "Greetings from Far Away." *New York Review of Books* (May 3, 1973) 16-18.

   Arthur Ford, Allen Spraggett with W.V. Rauscher
   ESP and Hypnosis, Susy Smith

   - *S:GBB* (1981) 251-261.

5. "What Hath Hoova Wrought?" *New York Review of Books* (May 16, 1974) 18-19.

   Uri, Andrija Puharich
   Arigo, John G. Fuller

   - *S:GBB* (1981) 275-288.

6. "Hocus-Pocus." *Washington Post Book World* (January 12, 1975) 3.

   The New Nonsense, Charles Fair
   Supersenses, Charles Paroti

   - *S:GBB* (1981) 289-291.

7. "Just Out of this World." *Washington Post Book World* (May 20, 1975) B4.

   Uri Geller: My Story, Uri Geller
   The Amazing Geller, Martin Ebon (Ed)

8. "Paranonsense." *New York Review of Books* (October 30, 1975) 14-15.

   Superminds, John Taylor
   The Magic of Uri Geller, The Amazing Randi

   - *S:GBB* (1981) 293-299.

9. "Strawberry Shortcut." *New York Review of Books* (December 11, 1975) 46-47.

   Powers of the Mind, Adam Smith

   - *S:GBB* (1981) 301-306.

10. "Supergull." *New York Review of Books* (March 17, 1977) 18-19.

    Mind-Reach, R. Targ and H. Puthoff
    The Search for Superman, J.L. Wilhelm

    - *S:GBB* (1981) 315-326.

11. "ESP at Random." *New York Review of Books* (July 14, 1977) 37.

    Learning to Use Extrasensory Perception, C. Tart

    - *S:GBB* (1981) 327-333.

12. "The Occult Underground." *Zetetic Scholar* **1**, 1 (1978) 51-53.

    The Occult Underground, J. Webb

13. "The Third Coming." *New York Review of Books* (January 26, 1978) 21-22.

    Close Encounters of the Third Kind, S. Spielberg
    The Hynek UFO Report, J.A. Hynek

    - *S:GBB* (1981) 347-359.
    - *TNIL* (1996) 244-254.

14. "Paper Bomber." *American Film* **3**, 8 (June 1978) 72-73.

    Four Arguments for the Elimination of Television, Jerry Mander

    - *S:GBB* (1981) 361-364.

15. *Skeptical Inquirer* **4**, 3 (Spring 1980) 60-62.

    Science and Parapsychology: A Critical Re-evaluation, C.E.M. Hansel

16. "Bottled Mischief." *Science Digest* (September 1981) 124.

    Polywater, Felix Franks

    - *W&W* (1989) 155-156.

17. "According to Hoyle." *Discover* **3**, 3 (March 1982) 12.

    Evolution from Space, F. Hoyle and C. Wickramasinghe

    - *O&S* (1983) 358-360.

18. *Vector* **6**, 2 (April 1982) 23-25.

    Theomatics, Jerry Lucas and Del Washbur

    - *O&S* (1983) 355-357.

19. "Boo." *Inquiry* (June 1982) 50-51.

   Poltergeist! A Study in Destructive Haunting, Colin Wilson

   - *O&S* (1983) 361-364.

20. "The Gribbin Effect." *Discover* **3**, 7 (July 1982) 10-11.

   The Jupiter Effect Reconsidered, John Gribbin and Stephen Plage-mann

   - *O&S* (1983) 365-368.

21. *Discover* (September 1982) 87-89.

   Abusing Science: The Case Against Creationism, P. Kitchner

   - *O&S* (1983) 372-374.

22. "Eysenck's Folly." *Discover* **3**, 10 (October 1982) 12.

   Astrology: Science and Superstition, Hans Eysneck and D.K.B. Nias

   - *O&S* (1983) 375-377.

23. *Skeptical Inquirer* **7**, 1 (Fall 1982) 64-65.

   Ordinary Daylight, A. Potok

   - *O&S* (1983) 378-379.

24. "Nuts about PK." *Nature* **300** (November 11, 1982) 119-120.

   Psychokinesis, John Randall

   - *NA* (1988) 179-181.

25. "Fools' Paradigms." *Free Inquiry* (Fall 1983) 46-47.

   Frames of Meaning, H. M. Collins and T. J. Pinch

   - *NA* (1988) 184-187.

26. "D. D. Home-Sweet-Home and Other Delusions." *Skeptical Inquirer* (Winter 1983-1984) 165-167.

   The Spiritualists, Ruth Brandon

   - *NA* (1988) 175-178.

27. "Of Crackpots and Clear Thinkers." *Science 84* (March 1984) 110, 112.

Dismantling the Universe, Richard Morris

- *W&W* (1989) 169-171.
- *WW&FL* (1996) 169-171.

28. "WAP, SAP, PAP, and FAP." *New York Review of Books* (May 8, 1986) 22-25.

The Anthropic Cosmological Principle, J. D. Barrow and F. J. Tipler

- *W&W* (1989) 218-228.
- *TNIL* (1996) 40-49.

29. "Isness is Her Business." *New York Review of Books* (April 9, 1987) 16-19.

Out on a Limb, Shirley MacLaine
Out on a Limb: ABC miniseries, Colin Higgins and Shirley MacLaine
Dancing in the Light, Shirley MacLaine
Ramtha, Steven Lee Weinberg

- *NA* (1988) 187-201.
- "The World According to Ram." *Utne Reader* (July/August 1988) 80-81..
- "Isness is Her Business: Shirley MacLaine." In *Not Necessarily the New Age*, edited by Robert Basil (Prometheus Books, 1988) 185-201.

30. "Propheteering Business." *Nature* (January 14, 1988) 125-126.

Bare-Faced Messiah, Russell Miller
L. Ron Hubbard, Bent Corydon and L. Ron Hubbard, Jr.

- *NA* (1988) 246-251.

31. "Bumps on the Head." *New York Review of Books* (March 17, 1988) 8-10.

Pseudo-Science and Society in the Nineteenth-Century America, A. Wrobel

- *OTWS* (1992) 130-137.
- *TNIL* (1996) 173-182.

32. "Seeing Stars." *New York Review of Books* (June 30, 1988) 43-45.

   Where's the Rest of Me, Ronald Reagan and R. G. Hubler
   Astrology for Teens, Joan Qigley
   Astrology for Adults, Joan Quigley
   Astrology for Parents of Children and Teenagers, Joan Quigley

   - *OTWS* (1992) 140-147.

33. "Linus Pauling from Chemist to Crank." *News and Observer* [Raleigh NC] (September 17, 1989).

   Linus Pauling: A Man and His Science, Anthony Serafini

34. "Not Daring to Doubt." *TLS: Times Literary Supplement* [London] (August 3 1990) 832.

   The Relentless Question, John Beloff

   - "The Search for the Holy Grail." *Free Inquiry* (Winter 1990/1991) 55-57.
   - *OTWS* (1992) 162-170.

35. "Excavating Cranks." *TLS: Times Literary Supplement* [London] (October 18, 1991) 6.

   Fantastic Archeology, Stephen Williams

   - *WW&FL* (1996) 228-230.

36. "Godfather of the New Age." *Washington Post Book World* (November 24, 1991) 1,9.

   A Fire in the Mind: The Life of Joseph Campbell, Stephen and Robin Larsen

   - *WW&FL* (1996) 231-233.

37. "A Book for Burning." *Nature* **360** (November 26, 1992) 396.

   Spontaneous Human Combustion, J. Randles and P. Hough

   - *WW&FL* (1996) 252-254.

38. "The New Age of Quackery." *Washington Post Book World* (March 17, 1996) 1,10.

   The Demon-Haunted World, Carl Sagan

   - *WJ* (2000) 125-133.

39. "When Superstition Displaces Science." *Nature* (January 30, 1997) 405-406.

   Behind the Crystal Ball, Anthony Aveni

   - *WJ* (2000) 142-145.

40. "The Magic of Therapeutic Touching." *Skeptical Inquirer* (November/December 2000) 48-49.

    Therapeutic Touch, Bela Scheiber and Carla Selby

    - "The Therapeutic Touch." In *UTB* (2003) 211-214.

41. "The Strange Case of Frank Jennings Tipler." *Skeptical Inquirer* **32**, 2 (March/April 2008) 57-59.

    The Physics of Christianity, Frank Tipler

    - *T&F* (2009) 181-185.

42. "Ann Coulter Takes on Darwin." *Skeptical Inquirer* **32**, 3 (May/June 2008) 54-55.

    Godless: The Church of Liberalism, Ann Coulter

    - *T&F* (2009) 3-8.

6.7   Published Correspondence

See also *Skeptical Inquirer* column chapter.

1. "Reich's Approval." *The Village Voice* (November 27, 1957) 4.

   Reply to a letter. He states that Reich had approved his chapter in *Fads*

2. "General Semantics." *Science* **128**, 3316 (July 18, 1958) 156.

   Rejoinder to Edd Doerr concerning Korzybski

3. "Evidence Disputed." *New York Times Magazine* (April 5, 1964) 68, 70.

   Encourages skepticism of claims of dermo-optical perception

4. "Was He Peeking?" *New York Review of Books* (July 28, 1966) 30.

   A reply to Brier and Pratt concerning Hansel.

   - *S:GBB* (1981) 223-225.

5. "Was He Peeking? (continued)." *New York Review of Books* (September 22, 1966) 29-30.

   A continued dialogue with Pratt.

   - *S:GBB* (1981) 227-228.

6. "A Spirited Exchange." *New York Review of Books* (November 1, 1973) 38.

   A reply to Allen Spraggett.

   - *S:GBB* (1981) 258-259.

7. "Trick or Treatment." *New York Review of Books* (July 18, 1974) 40-41.

   A reply to John Fuller concerning Arigo.

   - *S:GBB* (1981) 285.

8. "Pioneer." *New York Review of Books* (January 22, 1976) unknown.

   A reply to Adam Smith.

   - *S:GBB* (1981) 305-306.

9. *Scientific American* **234**, 1 (January 1976) 8.

   Rejoinder to Targ, Puthoff and Cole

   - *S:GBB* (1981) 88.

10. *Technology Review* **79**, 1 (October/November 1976) 2-3.

    Rejoinder to Puthoff and Targ

    - *S:GBB* (1981) 107-108.

11. "Psi and Science." *New York Review of Books* (October 13, 1977) 45.

    A reply to Charles Tart.

    - *S:GBB* (1981) 331-333.

12. *Humanist* (September/October 1977) 54.

    A rejoinder to a letter by Eldon A. Byrd.

    - *S:GBB* (1981) 177-178.

13. *Humanist* (January/February 1978) 2.

    A rejoinder to a letter by D. Scott Rogo.

    - *S:GBB* (1981) 146-147.

14. "Close Encounters." *New York Review of Books* (March 23, 1978) 53.

    A reply to B. Hopkins concerning Hynek.

    - *S:GBB* (1981) 356.

15. "Not a Mountebank." *New York Review of Books* (October 25, 1979) 52.

    A reply to Lynn Rose concerning Velikovsky.

    - *S:GBB* (1981) 386-387.

16. *New York Review of Books* (October 25, 1979) 53.

    A reply to D. Kline concerning Velikovsky.

    - *S:GBB* (1981) 388.

17. "Velikovsky's Deluge." *New York Review of Books* (March 6, 1980) 53.

    Another reply to Lynn Rose concerning Velikovsky.

    - *S:GBB* (1981) 389.

18. "Parapsychology: An Exchange." *New York Review of Books* (June 26, 1980) 50-51.

    A rejoinder to a letter of Beauregard, et al., on J.A. Wheeler's views.

    - *S:GBB* (1981) 199-201.

19. "Parapsychology & Physics." *New York Review of Books* (December 18, 1980) 68.

    A rejoinder to a rebuttal by Crussard and Bouvaist.

    - *S:GBB* (1981) 202-203.

20. *New York Review of Books* (December 18, 1980) 69.

    A reply J.B. Hasted.

    - *S:GBB* (1981) 204-205.

21. "Claims for ESP." *New York Review of Books* (February 19, 1981) 46.

    A reply to Charles Tart.

    - *S:GBB* (1981) 213-214.

22. "More on Ape Talk." *New York Review of Books* (April 2, 1981) 43.

    A reply to Washburn and Patterson.

23. *Science Digest* (January 1982) 12.

    Comment on an article on spontaneous combustion.

24. *Science Digest* (April 1982) 8.

    A reply to a review of *Science: Good, Bad, and Bogus.*

25. *Discover* (January 1983).

    A reply to Gauguelin.

    - *O&S* (1983) 377.

26. *Skeptical Inquirer* **8**, 2 (Winter 1983/1984) 187.

    Reply to Truzzi.

27. *The Sciences* [New York Academy of Sciences] (September/October 1984) 12,14.

    Reply to Louis Lasagna's critical review of the *Skeptical Inquirer.*

28. "Orwellian Possibilities." *The Sciences* [New York Academy of Sciences] (March/April 1985) 14.

    Reply to Louis Lasagna's rejoinder.

29. *Analog Science Fiction / Science Fact* (May 1985) 196-187.

    Strong reply to an article by "superpsychic" Alan Vaughn.

30. "The FAP Flop." *New York Review of Books* (December 4, 1996) 61.

   A two-word response to Frank Tipler's long letter: "I'm speechless."

31. *Skeptical Inquirer* **11**, 4 (Summer 1987) 424.

   Reply to McGirvey on EPR.

32. *Free Inquiry* (Spring 1988) 65.

   Rejoinder to Geier about Ray Palmer

   - *NA* (1988) 219-222.

33. "More Bumps on the Head." *New York Review of Books* (June 2, 1988) 41.

   Rejoinder to J. Winston about homeopathy.

   - *OTWS* (1992) 137-139.
   - *TNIL* (1996) 181.

34. "Paranormal Companionship." *New York Review of Books* (February 16, 1989) 45.

   Response to a review by I. Rosenfield of *The Oxford Companion to the Mind* concerning the inclusion of entries by Colin Wilson.

   - *OTWS* (1992) 155-156.

35. "Ghosts and Poltergeists." *New York Review of Books* (June 15, 1989) 458.

   Rejoinder to Colin Wilson about the above comments

   - *OTWS* (1992) 156-158.

36. "Questions for the Cosmos." *New York Times Magazine* (December 24, 1989) 4.

   Questions Jahn's controls of ESP experiments.

37. "*Discover*'s Advertisements." *Science* (February 2, 1990) 515.

   Explains why he disassociated himself from *Discover*.

38. "Reply to 'Unfinished Business'." *Skeptical Inquirer* **16**, 1 (Fall 1991) 88-89.

   A reply to Keith Harary's response about Russell Targ.

39. "The Shaver Mystery." *Fate* (November 1991) 83.

A reply to Jerome Clark concerning Ray Palmer's beliefs.

40. "Campbell Pro and Con." *Washington Post Book World* (December 29, 1991) 10,12.

A rejoinder to letters from Stephen and Robin Larsen and Terry Peay about Joseph Campbell

- *WW&FL* (1996) 234-236.

41. *Skeptical Briefs* **4**, 1 (March 1994) 13.

Reply about Stephen Barr.

42. "Extreme Credulity." *Skeptical Inquirer* **19**, 3 (May/June 1995) 29.

Response to an article by John Beloff.

43. *Skeptical Inquirer* **20**, 2 (March/April 1996) 64.

Rejoinder to Tort about John Beloff and Gardner's May/June 1995 letter.

44. *Washington Post Book World* (May 12, 1996) 14.

Rejoinder to letters about Sagan's book.

- *WJ* (2000) 133.

45. *The Skeptic* **5**, 3 (1997) 33.

Response to James French concerning Korzybski

46. *Saucer Smear* **45**, 3 (1998).

Request to be taken off the mailing list.

47. *Skeptical Inquirer* **30**, 3 (May/June 2006) 67.

Response to Douglas Hintzman about "Memory Wars"

48. "Gardner on Flew." *Skeptic* **19**, 4 (Winter 2006).

Response to Anthony Flew about Darwin's deism

SCIENCE AND THE UNKNOWABLE

MARTIN GARDNER

Existence, the preposterous miracle of existence! To whom has the world of opening day never come as an unbelievable sight? And to whom have the stars overhead and the hand and voice nearby never appeared as unutterably wonderful, totally beyond understanding? I know of no great thinker of any land or era who does not regard existence as the mystery of all mysteries.

—*John Archibald Wheeler*

One of the fundamental conflicts in philosophy, perhaps the most fundamental, is between those who believe that the universe open to our perception and exploration is all there is, and those who regard the universe we know as an extremely small part of an unthinkably vaster reality. These two views were taken by those two giants of ancient Greek philosophy, Plato and Aristotle. Plato, in his famous cave allegory, likened the world we experience to the shadows on the wall of a cave. To turn this into a mathematical metaphor, our universe is like a projection onto three-dimensional space of a much larger realm

---

Martin Gardner's "Science and the Unknowable" originally appeared in the *Skeptical Inquirer* 22, no. 6 (November/December 1998).

Figure 6.1: Column reprinted in *Science and Religion* (2003).

7

The *Skeptical Inquirer* Column

The name changed to "Notes of a Fringe Watcher" to broaden the scope, in vol. 11, no. 2, Winter 1986/1987. The final column appeared in vol. 26, no. 1, January/February 2002. There were 89 columns, which were collected into several books. Chapter / column correspondences are given

7.1   Chapters to Columns Index

- *The New Age: Notes of a Fringe-Watcher* (Prometheus, 1988).

1	Project Alpha	13-18	Summer 1983
2	Margaret Mead	19-24	Fall 1983
3	Magicians in the Psi Lab	25-31	Winter 1983/84
4	Shirley MacLaine	32-37	Spring 1984
5	Freud, Fliess, and Emma's Nose	38-43	Summer 1984
6	Koestler Money Down the Psi Drain	44-49	Fall 1984
7	Targ: From Puthoff to Blue	50-56	Winter 1984/85
8	The Relevance of Belief Systems	57-64	Spring 1985
9	Welcome to the Debunking Club	65-71	Summer 1985
10	The Great Stone Face	72-78	Fall 1985

Chapters 20 to 32 are book reviews and essays.

- *On the Wild Side* (Prometheus, 1992).

Chapters 17 to 32 are book reviews and essays.

- *Weird Water & Fuzzy Logic: More Notes of a Fringe Watcher* (Prometheus, 1996).

Chapters 17 to 46 are book reviews.

- *Did Adam and Eve Have Navels?* (Norton, 2000).

Chapter 26 is "The Wandering Jew," *Free Inquiry* (Summer 1995) pp. 31-33.

- *Are Universes Thicker than Blackberries?* (Norton, 2003).

The remaining 20 chapters are essays and reviews, several previously unpublished.

7.2 Other Books with Columns

1. Paul Kurtz (Ed.), *A Skeptic's Handbook of Parapsychology* (Prometheus, 1985).
   "Magicians in the Psi Lab: Many Misconceptions," pp. 351-356. (SI, Winter 1983/1984)

2. Kendrick Frazier (Ed.), *Science Confronts the Paranormal* (Prometheus, 1986).
   "The Great Stone Face and Other Mysteries," pp. 75-78. (SI, Fall 1985)
   "Lessons of a Landmark PK Hoax," pp. 166-169. (SI, Summer 1983)
   "Magicians in the Psi-Lab: Many Misconceptions," pp. 170-175. (SI, Winter 1983/1984)

3. Kendrick Frazier (Ed.), *The Hundredth Monkey* (Prometheus, 1991).
   "Psi Researchers' Inattention to Conjuring," pp. 162-166. (SI, Winter 1985/86)
   "The Obligation to Disclose Fraud," pp. 167-170. (SI, Spring 1988)
   "Water with Memory? The Dilution Affair," pp. 364-37.1 (SI, Winter 1989)
   "Science, Mysteries and the Quest for Evidence," pp. 372-375. (SI, Summer 1986)
   "Relativism in Science," pp. 376-380. (SI, Summer 1990)

4. Gero von Randow (Ed.), *Mein Paranormales Fahrrad und Andere Anlässe zur Skepsis, entdeckt im 'Skeptical Inquirer'* (Hamburg: Rowohlt, 1993). German translation by Anita Ehlers.
   "Mysteriöse Fingerübungen," pp. 99-106. (SI, Fall 1990)

5. Joe Nickell, Barry Karr, and Tom Genoni (Eds.), *The Outer Edge: Classic Investigations of the Paranormal* (Committee for Scientific Investigation of the Paranormal (CSICOP), 1996).
   "The Great Stone Face and Other Nonmysteries," pp. 64-67. (SI, Fall 1985)

6. Kendrick Frazier (Ed.), *Encounters with the Paranormal: Science, Knowledge, and Belief* (Prometheus Books, 1998).
   "Science vs. Beauty," pp. 49-56. (SI, Mar/Apr 1995)
   "Alan Sokal's Hilarious Hoax," pp. 88-93. (SI, Nov/Dec 1996)

"Heaven's Gate: The UFO Cult of Bo and Peep," pp. 146-151. (SI, Jul/Aug 1997)
"RMT: Repressed Memory Therapy," pp. 396-405. (SI, Summer 1993)

7. Sandra Fehl Tropp and Ann Pierson D'Angelo (Eds.), *Essays in Context* (Oxford University Press, 2001).
"The Great Samoan Hoax," pp. 705-711. (SI, Winter 1993); begins with a two page biographical sketch.

8. Paul Kurtz (Ed.), *Science and Religion: Are They Compatible?* (Prometheus, 2003).
"Science and the Unknowable," pp. 321-329. (SI, Nov/Dec 1998)

9. Keith M. Parsons (Ed.), *The Science Wars* (Prometheus, 2003).
"Science or Beauty," pp. 307-314. (SI, Mar/Apr 1995)

10. Kendrick Frazier (Ed.), *Science Under Siege* (Prometheus, 2009).
"The Memory Wars [Part 1 only]," pp. 133-138. (SI, January/February 2006)
" 'Dr.' Bearden's Vacuum Energy," pp. 311-314. (SI, January/February 2007)

7.3   Published Correspondence

1. *Skeptical Inquirer* **8**, 2 (Winter 1983-1984) p. 187.
Reply to Truzzi (related to Summer 1983).

2. *Skeptical Inquirer* **9**, 3 (Spring 1985) pp. 299-300.
Reply to Beloff (related to Fall 1984).
   • *NA*, 1988, pp. 48-49.

3. *Skeptical Inquirer* **10**, 1 (Fall 1985) pp. 91-92.
On science and Christians (related to Spring 1985).
   • *NA*, 1988, pp. 61-64.

4. *Skeptical Inquirer* **10**, 4 (Summer 1986) pp. 371-373.
Reply to Beloff about Homes (related to Winter 1985/1986).
   • *NA*, 1988, pp. 85-88.

5. *Skeptical Inquirer* **11**, 3 (Spring 1987) p. 314.
Reply to Gillett about Gold (related to Fall 1986).
   • *NA*, 1988, pp. 106-108.

6. *Skeptical Inquirer* **14**, 4 (Summer 1990) p. 440.
   Reply to Fulkerson about Urantia (related to Winter 1990).

7. *Skeptical Inquirer* **15**, 2 (Winter 1991) pp. 212-213.
   Reply to Gregory about relativism (related to Summer 1990).

   - *OTWS*, 1992, pp. 82-84.

8. *Skeptical Inquirer* **15**, 4 (Summer 1991) p. 443.
   Reply Edwards and Lippey about relativism (related to Summer 1990).

   - *OTWS*, 1992, pp. 84-86.

9. *Skeptical Inquirer* **15**, 4 (Summer 1991) p. 441.
   Reply Poul Anderson on Tipler (related to Winter 1991).

   - *OTWS*, 1992, pp. 100-101.

10. *Skeptical Inquirer* **16**, 2 (Winter 1992) p. 202.
    Reply to Mims (related to Summer 1991).

    - *OTWS*, 1992, pp. 115-117.

11. *Skeptical Inquirer* **16**, 3 (Spring 1992) p. 332.
    Corrects errors regarding Urantia (related to Winter 1990).

12. *Skeptical Inquirer* **16**, 4 (Summer 1992) pp. 441-442.
    Reply to Bunge and Gilkey on probability problems (related to Winter 1992).

    - *WW&FL*, 1996, pp. 19-21.

13. *Skeptical Inquirer* **17**, 4 (Summer 1993) pp. 439-440.
    Rejoinder to Cole, Schwartz, and Earley about Mead and Samoa, in the "Follow Up" section (related to Winter 1993).

    - *WW&FL*, 1996, pp. 43-50.

14. *Skeptical Inquirer* **18**, 1 (Fall 1993) pp. 106-107.
    Rejoinder on E-prime (related to Spring 1993).

    - *WW&FL*, 1996, pp. 57-62.

15. *Skeptical Inquirer* **18**, 2 (Winter 1994) pp. 214-215.
    Rejoinder to Karon, Shapiro, and Mims about RMT (related to Summer 1993).

    - *WW&FL*, 1996, pp. 70-73.

16. *Skeptical Inquirer* **18**, 4 (Summer 1994) p. 442.
    Reply to Shapiro's rejoinder about RMT (related to Summer 1993).
    - *WW&FL*, 1996, pp. 73-4.

17. *Skeptical Inquirer* **21**, 2 (March/April 1997) p. 62.
    Rejoinder to Nexon and Levine about Sokal (related to November/December 1996).
    - *Adam&Eve*, 2000, p. 152.

18. "Responding to Puthoff: Zero-Point Energy," *Skeptical Inquirer* **22**, 5 (September/October 1998) pp. 61-62.
    Response in the "Follow-Up" section (related to May/June 1998).
    - *Adam&Eve*, 2000, pp. 68-69.

19. *Skeptical Inquirer* **23**, 1 (January/February 1999) p. 66.
    Two rejoinders about Temple University (related to September/October 1998).
    - *Adam&Eve*, 2000, pp. 229-230.

20. *Skeptical Inquirer* **23**, 2 (March/April 1999) p. 67.
    A further rejoinder about Temple University (related to September/October 1998).
    - *Adam&Eve*, 2000, p. 230. (Beginning with "It's not easy ...".)

21. *Skeptical Inquirer* **24**, 6 (November/December 2000) p. 67.
    Terse response about Krishnamurti (related to July/August 2000).

22. *Skeptical Inquirer* **25**, 6 (November/December 2001) p. 72.
    Terse response to many letters about Popper (related to May/June 2001).
    - *UTB*, 2003, pp. 17-18.

23. *Skeptical Inquirer* **26**, 3 (May/June 2002) p. 61.
    Response in the "Follow-Up" section to DeWitt about multiverses (related to Mar/Apr 2002).
    - *UTB*, 2003, pp. 10-11.

7.4   Columns to Chapters Index

Vol/No		Issue	Year	Pages	Book	Chap
Original column					Book appearance	
Vol/No		Issue	Year	Pages	Book	Chap
7	4	Summer	1983	16-19	NEW AGE	1
8	1	Fall	1983	13-16	NEW AGE	2
8	2	Winter	1983/84	111-116	NEW AGE	3
8	3	Spring	1984	214-218	NEW AGE	4
8	4	Summer	1984	302-304	NEW AGE	5
9	1	Fall	1984	13-16	NEW AGE	6
9	2	Winter	1984/85	118-121	NEW AGE	7
9	3	Spring	1985	213-216	NEW AGE	8
9	4	Summer	1985	319-32	NEW AGE	9
10	1	Fall	1985	14-18	NEW AGE	10
10	2	Winter	1985/86	116-120	NEW AGE	11
10	3	Spring	1986	202-205	NEW AGE	12
10	4	Summer	1986	303-306	NEW AGE	13
Column title changed to "Notes of a Fringe Watcher"						
11	1	Fall	1986	21-24	NEW AGE	14
11	2	Winter	1986/87	128-131	NEW AGE	15
11	3	Spring	1987	236-238	NEW AGE	16
11	4	Summer	1987	337-340	NEW AGE	17
12	1	Fall	1987	29-32	NEW AGE	18
12	2	Winter	1987/99	128-131	NEW AGE	19
12	3	Spring	1988	240-243	WILD SIDE	1
12	4	Summer	1988	355-358	WILD SIDE	2
13	1	Fall	1988	26-30	WILD SIDE	3
13	2	Winter	1989	132-141*	WILD SIDE	4
13	3	Spring	1989	252-256	WILD SIDE	5
13	4	Summer	1989	357-361	WILD SIDE	6
14	1	Fall	1989	16-20	WILD SIDE	7
14	2	Winter	1990	124-129	WILD SIDE	8
14	3	Spring	1990	245-249	WILD SIDE	9
14	4	Summer	1990	353-357	WILD SIDE	10
15	1	Fall	1990	30-34	WILD SIDE	11
15	2	Winter	1991	128-132	WILD SIDE	12
15	3	Spring	1991	242-246	WILD SIDE	13
15	4	Summer	1991	355-361	WILD SIDE	15

Original column				Book appearance	
Vol/No		Issue	Year	Pages	Book	Chap
16	1	Fall	1991	27-30	WILD SIDE	14
16	2	Winter	1992	129-132	WEIRD WATER	1
16	3	Spring	1992	244-248	WILD SIDE	16
16	4	Summer	1992	357-361	WEIRD WATER	2
17	1	Fall	1992	17-23	WEIRD WATER	3
17	2	Winter	1993	131-133	WEIRD WATER	4
17	3	Spring	1993	261-266	WEIRD WATER	5
17	4	Summer	1993	370-375	WEIRD WATER	6
18	1	Fall	1993	15-20	WEIRD WATER	7
18	2	Winter	1994	126-130	WEIRD WATER	8
18	3	Spring	1994	243-247	WEIRD WATER	9
18	4	Summer	1994	356-9, 362	WEIRD WATER	10
18	5	Fall	1994	464-470	WEIRD WATER	11
19	1	Jan/Feb	1995	14-17	WEIRD WATER	12
19	2	Mar/Apr	1995	14-16, 55	WEIRD WATER	13
19	3	May/Jun	1995	9-11, 54	WEIRD WATER	14
19	4	Jul/Aug	1995	3-5, 55	WEIRD WATER	15
19	5	Sept/Oct	1995	9-11, 56	WEIRD WATER	16
19	6	Nov/Dec	1995	10-12, 56	ADAM & EVE	10
20	1	Jan/Feb	1996	7-9	ADAM & EVE	11
20	2	Mar/Apr	1996	12-15	ADAM & EVE	17
20	3	May/Jun	1996	8-10	ADAM & EVE	5
20	4	Jul/Aug	1996	9, 11-12	ADAM & EVE	20
20	5	Sept/Oct	1996	13-16	ADAM & EVE	22
20	6	Nov/Dec	1996	14-6	ADAM & EVE	14
21	1	Jan/Feb	1997	14-17	ADAM & EVE	12
21	2	Mar/Apr	1997	16-18, 57	ADAM & EVE	23
21	3	May/Jun	1997	14-15, 54	ADAM & EVE	18
21	4	Jul/Aug	1997	15-17*	ADAM & EVE	19
21	5	Sept/Oct	1997	16-17, 58	ADAM & EVE	24
21	6	Nov/Dec	1997	17-20	ADAM & EVE	2
22	1	Jan/Feb	1998	14-16	ADAM & EVE	13
22	2	Mar/Apr	1998	12-14	ADAM & EVE	1
22	3	May/Jun	1998	13-15, 60	ADAM & EVE	6
22	4	Jul/Aug	1998	16-19	ADAM & EVE	3
22	5	Sept/Oct	1998	14-17	ADAM & EVE	21
22	6	Nov/Dec	1998	20-23	ADAM & EVE	28

Original column				Book appearance	
Vol/No	Issue	Year	Pages	Book	Chap
23	1	Jan/Feb	1999	12-14	ADAM & EVE	15
23	2	Mar/Apr	1999	15-17	ADAM & EVE	8
23	3	May/Jun	1999	13-15	ADAM & EVE	9
23	4	Jul/Aug	1999	8-9, 12-13	ADAM & EVE	25
23	5	Sept/Oct	1999	13-15	ADAM & EVE	16
23	6	Nov/Dec	1999	13-15	ADAM & EVE	4
24	1	Jan/Feb	2000	9, 12-14	ADAM & EVE	27
24	2	Mar/Apr	2000	12-13, 30	BLACKBERRIES	13
24	3	May/Jun	2000	9, 12-14	ADAM & EVE	7
24	4	Jul/Aug	2000	20-23	BLACKBERRIES	15
24	5	Sept/Oct	2000	14-16	BLACKBERRIES	22
24	6	Nov/Dec	2000	12-14	BLACKBERRIES	23
25	1	Jan/Feb	2001	17-19	BLACKBERRIES	24
25	2	Mar/Apr	2001	12-14	BLACKBERRIES	25
25	3	May/Jun	2001	17-19	BLACKBERRIES	27
25	4	Jul/Aug	2001	13-14, 72	BLACKBERRIES	2
25	5	Sept/Oct	2001	14-16	BLACKBERRIES	1
25	6	Nov/Dec	2001	16-18	BLACKBERRIES	17
26	1	Jan/Feb	2002	12-14	BLACKBERRIES	4
31	1	Jan/Feb	2007	18-19	†	
33	5	Sept/Oct	2009	22-23	‡	
33	6	Nov/Dec	2009	22	††	
34	1	Jan/Feb	2010	16-17	‡‡	
34	2	Mar/Apr	2010	19-20	†††	
34	5	Sept/Oct	2010	10-11	‡‡‡	

Notes:

* Not labeled as the column, but was used instead as an unlabeled article in a "Special Report."

† " 'Dr.' Bearden's Vacuum Energy"

‡ "Bobby Fischer: Genius or Idiot"

†† "Bill Maher: Crank and Comic"

‡‡ "Thomas Gold: Is the Origin of Oil Nonbiological"

††† "James Arthur Ray: New Age Guru and Sweat Lodge Culprit"

‡‡‡ "Swedenborg and Dr. Oz"

Feb-20-36
The Daily Maroon

...Editorial...

They Who Think They Think Should Think Again

A college student is sooner or later inevitably faced with a choice of two alternative programs of mental action. On the one hand he can adopt a temporary scepticism to be discarded when he has advanced sufficiently into his field of concentration to feel justified in forming a permanent set of answers to the problems of that field. On the other hand he may dispense with this intermediary of sceptisism, and begin immediately to acquire and defend a system of dogmas.

The student who chooses to adopt the first policy does not hesitate to formulate a set of answers—but the formulation is recognized at the outset as tentative. He is aware of the possibility that his answers are wrong. Consequently as he advances in his acquisition of knowledge, he can change and modify his answers with relative ease. Ultimately he arrives at a set which is intellectually satisfying and which he feels adequate to defend.

The student who pursues the second policy never formulates a set of tentative answers. His original set is considered final. Usually it consists of answers which are thrown to him by the first instructor or set of instructors for whom he has great admiration.

He mirrors these instructors.

He doubts only what he is told by them to doubt.

His faith in the immutability of their answers is overpowering. Their answers form a scaffolding on which he hangs all his subsequent thinking. Modifications of that scaffolding are made only with increasing reluctance.

American collegiates were once characterized by the statement that they preferred being told what to think to being told to think.

And students who are told what to think are too often the students who think they do the most thinking.   —M. G.

Figure 7.1: University of Chicago's newspaper *The Daily Maroon*.

Chapter 8

Philosophy of Science

Recall that Martin Gardner's studies at the University of Chicago centered on the Philosophy of Science. His interest was less academic and more personal. He studied philosophy so that he "would know what to believe." After a decade of soul-searching, in a 1940 interview he gave the principle he held for the rest of his life (as expressed by Lord Dunsany): "Man is a small thing, and the night is large and full of wonder." After the war he studied with Rudolph Carnap, which led to his essay "Order and Surprise" (1950). It describes the deep sense of wonder that the natural world impresses upon philosophers and scientists. He described himself as a "mysterian" since he believed that even though Science will continue to resolve open problems, behind every mystery we will find another mystery to vex us.

He was a student of Carnap and, a couple of years later, Carnap allowed him to create transcripts of recordings of his lectures, which were in need of editing. Even though Carnap's name is on the book, Gardner largely wrote the text.

8.1 Books

Gardner published many eclectic anthologies that spanned categories. I have decided to list them here, as the Philosophy of Science was a major thread through them.

1. *Order & Surprise* (Prometheus, 1983).

   This book anthologizes most of Gardner's essays and reviews up to 1984 that had not appeared in *Science:Good, Bad, and Bogus*. Many of the articles are on philosophy. The rest are on diverse topics: literature, math, art, Oz, science, and fringe science.

2. *Gardner's Whys and Wherefores* (Chicago Press, 1989).

   This book anthologizes many of Gardner's essays and reviews that appeared up to 1989 on philosophy, literature, mathematics, and science.

   - *Gardner's Whys and Wherefores* (Prometheus Books, 1999). Contains a new "Preface to the Paperback Edition."

3. *The Night is Large: Collected Essays 1938-1995* (St. Martin's Press, 1996).

   This book anthologizes many of Gardner's best essays and is organized into seven parts: Physical Science, Social Science, Pseudoscience, Mathematics, the Arts, Philosophy, and Religion.

4. *From the Wandering Jew to William F. Buckley Jr.* (Prometheus, 2000).

   This book anthologizes many of Gardner's essays and reviews that appeared up to 2000 on religion, literature, mathematics, and science.

5. *The Jinn from Hyperspace* (Prometheus, 2008).

   This book anthologizes many of Gardner's essays and reviews that appeared up to 2008 on science, literature, and mathematics.

6. *When You Were a Tadpole and I was a Fish: and Other Speculations about This and That* (Hill and Wang, 2009).

   This book anthologizes many of Gardner's essays and reviews that appeared up to 2009 on science, literature, and mathematics.

8.2 Books Edited

1. Rudolf Carnap, *Philosophical Foundations of Physics* (Basic, 1966).
   - Rudolf Carnap, *Introduction to the Philosophy of Science* (Basic, 1974).

     With a new foreword.

- Rudolf Carnap, *Introduction to the Philosophy of Science* (Dover, 1995).

  With a new foreword.

- "Rudolf Carnap, Philosopher of Science." In *UTB* (2003) 19-22. The foreword to the Dover edition.

8.3 Book Introductions

1. H. G. Wells, *Anticipations* (Dover, 1999) iii-xii.

   - *WJ* (2000) 250-262.

     The last third of chapter 23.

8.4 Articles

1. "Hermit Scholars." *The Daily Maroon* [University of Chicago] (March 14, 1934) 2.

   This appeared as an editorial; on this date "Martin Gardiner" [sic] was added to the masthead of this student newspaper.

2. "They Who Think They Think Should Think Again." *Daily Maroon* [University of Chicago] (February 20, 1936) 3.

3. "Skepticism and Ray Ellinwood." *Daily Maroon* [University of Chicago] (December 2, 1936) 3.

4. "The Fire Burning." *The Daily Maroon* [University of Chicago] (May 19, 1937) 1-2.

5. "The Strange Case of Robert Maynard Hutchins." *University Review* [University of Kansas City] **5**, 2 (Winter 1938) 84-89.

   - *O&S* (1983) 13-28.
   - *TNIL* (1996) 509-524.

6. "Order and Surprise." *Philosophy of Science Quarterly* **17**, 1 (January 1950) 109-117.

   - *O&S* (1983) 57-67.

7. "Mathematics and the Folkways." *Journal of Philosophy* **47**, 7 (March 30, 1950) 177-186.

   - *O&S* (1983) 68-80.
   - *TNIL* (1996) 257-270.

8. "Beyond Cultural Relativism." *International Journal of Ethics* **61**, 1 (October 1950) 38-40.

    - *O&S* (1983) 86-97.
    - *TNIL* (1996) 149-161.

9. "Is Nature Ambidextrous." *Philosophy and Phenomenological Research* **13**, 2 (December 1952) 200-211.

    - *O&S* (1983) 103-114.

10. "Bourgeois Idealism in Soviet Nuclear Physics." *Yale Review: New Series* **43** (1954) 386-399.

    - *S:GBB* (1981) 15-25.

11. "Logical Paradoxes." *Antioch Review* **23** (1963) 172-178.

    - *O&S* (1983) 157-163.

12. "Can Time Go Backward?" *Scientific American* **216** (January 1967) 98-108.

    - "Can Time Go Backward?" In *Readings in the Physical Sciences: Volume 3* (W. H. Freeman, [1968]) 986-993. (Offprint 309)
    - "Can Time Go Backward?" In *The Project Physics Course: Reader 6* (Holt, Rinehart and Winston, 1971) 193-200.
    - "UTB." *Cacumen* [Spanish] **3**, 3 (2003) 29-45.

13. "The Fall of Parity." In *The Project Physics Course: Reader 6* (Holt, Rinehart and Winston, 1971) 175-191.

    Chapter from *The Ambidextrous Universe*

14. "Quantum Weirdness." *Discover* **3**, 10 (October 1982) 69-76.

    Won the 1983 Science-Writing Award of the American Institute of Physics.

    - *O&S* (1983) 194-204.
    - *TNIL* (1996) 22-31.

15. "Anti-Science: The Strange Case of Paul Feyerabend." *Free Inquiry* (Winter 1982/1983) 32-34.

    - *O&S* (1983) 205-210.

16. "Feyerabend and Science." *Free Inquiry* (Summer 1983) 59.

    This is a rejoinder to Feyerabend's reply to the Winter 1982/1983 article.

- "Anti-Science: The Strange Case of Paul Feyerabend." In *On the Barricades: Religion and Free Inquiry in Conflict*, edited by R. Basil, et al (Prometheus, 1989) 268-279.

  Combines the text from the Winter 1982/1983 and Summer 1983 issues.

17. "Those Mindless Machines." *Washington Post* (May 25, 1997) C1, C5.

    - *GW* (2001) 91-95.

18. "Is Reuben Hersh 'Out There'?" *EMS* [European Mathematical Society] *Newsletter* (June 2009) 23-24.

    - *T&F* (2009) 124-128.

8.5   Book Reviews

1. "Does Two Plus Two Equal Four." *New Leader* (January 14, 1950) 9.

   The Science of Culture, Leslie White

   - *O&S* (1983) 238-242.

2. "Relativity: Hope Chest or Pandora's Box?" *New Leader* **33**, 32 (August 12, 1950) 26.

   Relativity - A Richer Truth, P. Frank

   - *O&S* (1983) 243-244.

3. "Lord Russell in Photographs." *New Leader* (September 16, 1950) 21.

   Bertrand Russell O.M., Harry Leggett

4. "Einstein and Relativity." *New Leader* **34**, 6 (February 5, 1951) 23-24.

   Albert Einstein: Philosopher Scientist, Edited by P.A. Schlipp

   - *O&S* (1983) 245-246.

5. "Philosophy as Poetic Speculation." *New Leader* **34**, 24 (June 11, 1951) 23-24.

   The Rise of Scientific Philosophy, H. Reichenbachs

   - *O&S* (1983) 247-248.

6. "Lord Russell Views the Future." *New Leader* **34**, 33 (August 13, 1951) 19.

   The Impact of Science on Society, Bertrand Russell

7. "World Science (Ltd.)." *New Leader* **34**, 29 (August 16, 1951) 26.

   Science: Its Method and Its Philosophy, G.B. Brown

8. *Philosophy and Phenomenological Research* (September 1953) 135-136.

   Problems of Life, Ludwig von Bertalanffy

9. *New York Herald–Tribune: Books* (September 2, 1962) 9.

   Fact and Fiction, Bertrand Russell

10. "I Think it Thinks, Ergo Who Is?" *New York Herald Tribune's Book Week* (June 28, 1964) 6,17.

    Analytic Engine, Jeremy Bernstein
    God and Golem Inc., Norbert Weiner

    - *O&S* (1983) 255-25.

11. "Methods and Madness." *New York Herald Tribune's Book Week* (October 2, 1966) 14.

    Ideas in Conflict, Theodore Gordon

    - *S:GBB* (1981) 233-235.

12. "Up From Adam." *New York Herald Tribune's Book Week* (April 23, 1967) 3,13.

    The Biology of Ultimate Concern, T. Dobzhansky

    - *O&S* (1983) 259-262.

13. "Explaining the World." *Commentary* (October 1967) 100-102.

    A Comprehensible World, Jeremy Bernstein

    - *O&S* (1983) 263-265.

14. *Library of Science* (1967). Advertising brochure.

    Mind and Cosmos, Edited by R.G. Colodny

15. "Can Computers Ever Learn How to be People." *Washington Post Book World* (January 23, 1972) 12.

    What Computers Can't Do: A Critique of Human Reason, H.L. Dreyfus

    - *O&S* (1983) 277-279.

16. "Arthur Koestler: Neoplatonism Rides Again." *World* **1**, 3 (August 1, 1972) 67-69.

    The Roots of Coincidence, Arthur Koestler

    - *S:GBB* (1981) 241-249.

17. "A Subjective Approach to the Paradoxes of Physics." *Newsday: Ideas* (May 27, 1979) 20,17.

    The Dancing Wu Li Masters, Gary Zukav
    Lifetide, Lyall Watson

    - *S:GBB* (1981) 375-378.

18. "Cracking Creating." *New York Times Book Review* (January 3, 1982) 6-7.

    The Minds Best Work, D.N. Perkins

    - *O&S* (1983) 352-354.

19. *Discover* **4**, 4 (April 1983) 92-93.

    Betrayers of the Truth, W. Broad and N. Wade

    - *NA* (1988) 182-183.

20. "The Value of the Variable, or To Be Is to Quine." *Boston Sunday Globe* (July 7, 1985) B10, B12.

    The Time of My Life, W. V. Quine

    - *W&W* (1989) 211-214.
    - *TNIL* (1989) 491-494.

21. "The Curious Mind of Allan Bloom." *Education and Society* **1**, 1 (Spring 1988) 29-32.

    The Closing of the American Mind, Allan Bloom

    - *W&W* (1989) 256-261.
    - *TNIL* (1996) 450-456.

22. "Searching for the Meaning of It All." *News and Observer* [Raleigh, NC] (October 30, 1988) 5.

    Three Scientists and Their Gods, Robert Wright

    - *WW&FL* (1996) 172-174.

23. "And Bang! Went the Universe." *News and Observer* [Raleigh, NC] (September 3, 1989).

    Reading the Mind of God, James Trefil

    - *WW&FL* (1996) 201-202.

24. "A World That is Rediscovering Itself." *News and Observer* [Raleigh, NC] (December 2, 1989).

    Disappearing through the Skylight, O. B. Hardison Jr.

    - *WW&FL* (1996) 203-205.

25. "Heisenberg: Yin vs. Yang." *Dimensions* **7**, 1 (1993).

    Uncertainty: The Life and Science of Werner Heisenberg, David C. Cassidy

    - *WW&FL* (1996) 68-77.

26. *Los Angeles Times Book Review* (May 18, 1997) 5.

    Achilles in the Quantum Univers, Richard Morris
    The Invention of Infinity, J. V. Field

27. "Amazing Grace." *Los Angeles Times Book Review* (February 1, 1988) 5.

    The Universe and the Teacup, K. C. Cole

    - *GW* (2001) 281-284.

28. "Maybe." *Los Angeles Times Book Review* (November 29, 1988) 8-9.

    Probability 1, Amir D. Aczel

    - *GW* (2001) 295-299.

29. "True Confessions." *Los Angeles Times Book Review* (August 6, 2000) 4-5.

    Papal Sin, Garry Wills

    - "The Strange Case of Gary Willis." In *UTB* (2003) 85-93. Complete uncut version.

8.6   Published Correspondence

1. *Scientific American* (April 1956) 20.

   Shows Sextus Empiricus anticipated Wittgenstein.

2. "A New Prediction Paradox." *British Journal for the Philosophy of Science* **13**, 49 (May 1962) 51.

   Karl Popper gives a favorable reply.

3. "John Dewey." *New York Times Book Review* (November 1, 1964) 30.

   Notes William James and C. S. Pierce anticipated Dewey's pragmatism.

4. *Scientific American* (September 1965) 10,12.

   Lengthy and varied response about travel in gravity tubes.

5. *Physics Today* **21**, 3 (March 1968) 15.

   Defends the philosophy of realism.

6. "Not Freud's Discovery." *New York Review of Books* (June 12, 1975) 45-46.

   Discusses Pre-Freudian theories

7. "On Einstein, Newton and the Art of Playwriting." *New York Times* (September 4, 1977) D4.

   Corrects an essay by Robert Brustein, "Drama in the Age of Einstein."

8. "Facing Up to Realism." *New York Review of Books* (January 21, 1982) 68.

   Defends realism in a rejoinder to Robert Farrell.

   - *O&S* (1983) 348-349.
   - *TNIL* (1996) 290-291.

9. *Hypotheses: Neo-Aristotelian Analysis* 3 (Fall 1992) 11.

   Discussion of McKeon, Adler and Wells.

10. "Prediction Paradox." *Mathematical Intelligencer* **16**, 2 (1994) 3.

    Replies to a discussion about prediction paradoxes.

11. *Commentary* (May 1998) 7.

    Questions the theological agenda of Berlinski.

It was weeks after my conversion before I ever attended a revival meeting or heard a real Christian preach. I had been reading literature from the freethought press and priding myself on my ability to see through religious sham, when one day I really began to find out what the Gospel was about. The series of circumstances that led me to do this were so complicated that I don't believe I ever fully understood just who or what it was that led me to Christ. A great many things happened along about this time and as I look back over it I cannot help but marvel at the part each little detail played in preparing my mind for that which was to follow. Suffice it to say that I started to pray and as I learned more about Christianity and became convinced of its truth, I began to wonder if I was saved. I prayed for the experience of conversion and it was not until I found out just what conversion was from some Moody Colportage tracts, that I realized that I was already a ransomed soul. All this happened several years ago. Exactly how or why it happened I do not know. But nevertheless it *did* happen and I have never ceased to praise Him for it.

—from a student in the University of Chicago.

Figure 8.1: Letter written before his dissolution with fundamentalism. He acknowledged writing the letter, though later he wished he had not.

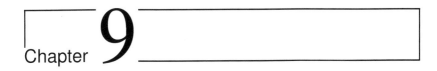

Chapter **9**

Philosophy and Theology

Martin Gardner experienced a dissolution of his fundamentalist Christian faith while at the University of Chicago. It did not occur suddenly; the slow decay is detailed in his novel *The Flight of Peter Fromm* (see Fiction chapter). He even enrolled for a year (on a scholarship) at the affiliated Chicago Theological Seminary. By that time (1936-1937) he had lost his faith in religion but not God. His magnum opus is *The Whys of a Philosophical Scrivener*, which he described as a "confessional," wherein he justifies his beliefs. He had little patience with beliefs and people he thought were ridiculous.

He should have written a textbook on Philosophy inasmuch as until his death at 95 he could discourse impromptu on any topic in the history of Western Philosophy. His facility with that material is seen throughout these writings.

9.1   Books

1. *The Whys of a Philosophical Scrivener* (William Morrow, 1983).

   - *The Whys of a Philosophical Scrivener* (Quill, 1983).
   - *The Whys of a Philosophical Scrivener* (St. Martin's Griffin, 1999).

   Contains a lengthy new "Postscript."

153

2. *The Healing Revelations of Mary Baker Eddy: The Rise and Fall of Christian Science* (Prometheus Books, 1993).

   - *The Life of Mary Baker Eddy: The Rise and Fall of Christian Science* (Battered Silicon Dispatch Box, 2008).

   Contains a new preface, corrections, and a small addition.

3. *Urantia: The Great Cult Mystery* (Prometheus Books, 1995).

   - *Urantia: The Great Cult Mystery* (Prometheus Books, 2008).

   Contains a new postscript.

9.2 Book Introductions

1. Steve Allen, *Steve Allen on the Bible, Religion, and Morality* (Prometheus, 1991) xi-xvi.

9.3 Articles

1. "A Sophomore Speaks." *Revelation* (September 1934) 340, 366.
2. "Glossolalia." *Free Inquiry* (Spring 1989) 46-48.

   - *OTWS* (1992) 173-181.
   - "Glossolalia." *Indian Skeptic* 5, 12 (April 1993) 10-14.

3. "Keeping up with Oral Roberts' City of Faith." *Time-News* [Hendersonville, NC] (October 21, 1989).

   - "Keeping up with Oral Roberts's City of Faith." *Free Inquiry* (Winter 1989/1990) 49.
   - *OTWS* (1992) 188-191.

4. "Some Fatherly Advice to Tammy Faye." *Free Inquiry* (Summer 1990) 45-46.

   - *OTWS* (1992) 192-195.

5. "New Thought, Unity, and Ella Wheeler Wilcox." *Hypotheses: Neo-Aristotelian Analysis* 5 (Spring 1993) 8-11.

   Excerpted from *The Healing Revelations of Mary Baker Eddy*; an additional excerpt in the editor's column on page 1.

   - *T&F* (2009) 49-74.

6. "Who are the Branch Davidians." *Times-News* [Hendersonville, NC] (April 27, 1993) 10A.

7. "Oral Roberts on Jim Bakker." *Free Inquiry* (Summer 1994) 15.

   • *WJ* (2000) 29-31.

8. "The Wandering Jew and the Second Coming." *Free Inquiry* (Summer 1995) 31-33.

   • *TNIL* (1996) 525-532.
   • *Adam&Eve* (2000) 274-287.
   • *WJ* (2000) 11-24.
   • "The Wandering Jew and the Second Coming." In *The Portable Atheist*, edited by Christopher Hutchins (Third Da Capo Press, 2007) 211-217.

9. "The Incredible Flimflams of Margaret Rowen, Part 1: Seventh Day Adventists and the Second Coming." *Free Inquiry* (Spring 1996) 36-40.

   • *WJ* (2000) 37-50.

10. "The Incredible Flimflams of Margaret Rowen, Part 2: The Sad Saga of Dr. Bert Fullmer." *Free Inquiry* (Fall 1996) 38-42.

    • *WJ* (2000) 51-65.

11. "The Incredible Flimflams of Margaret Rowen, Part 3: The Comic Pratfalls of Robert Reid[t]." *Free Inquiry* (Spring 1997) 53-56.

    • *WJ* (2000) 66-76.

12. "It's Official: Adler Finally Converts." *Free Inquiry* (Fall 2000) 9.

    Short piece that includes his "The Road to Rome" letter.

13. "Surprise." In *This Is My Best*, edited by Retha Powers and Kathy Kiernan (Quality Paperback Book Club, 2004) 146-165.

    Chapter 20 of *Whys* with a new preamble by Gardner

    • "Surprise." In *This Is My Best*, edited by Retha Powers and Kathy Kiernan (Chronicle Books, 2005) 146-165.

14. "The Comic Pratfalls of Richard Roberts." *Skeptical Inquirer* **32**, 2 (March/April 2008) 5-6.

    • *T&F* (2009) 186-189.

15. "Why I Am Not an Atheist." In *When You were a Tadpole I was a Fish* (Hill and Wang, 2009) 190-206.

    Chapter 13 of *Whys*

9.4 Book Reviews

1. *International Journal of Ethics* **49**, 4 (July 1939) 501-502.

   The American State University, N. Foester

2. *International Journal of Ethics* **49**, 4 (July 1939) 503.

   Plato Today, R.H.S. Crossman

3. *International Journal of Ethics* **49**, 4 (July 1939) 508.

   Perish the Jew, David Goldberg
   Twenty Centuries of Jewish Thought, A. Lichtigfeld

4. *International Journal of Ethics* **50**, 2 (January 1940) 235-236.

   A Christian Look at the Jewish Question, Jacques Maritain

   - *O&S* (1983) 219-220.

5. *International Journal of Ethics* **58**, 2 (January 1948) 144-146.

   The Kafka Problem, Edited by Angel Fiona
   Kafka's Prayer, P. Goodman

   - *O&S* (1983) 221-224.

6. *International Journal of Ethics* **58**, 4 (July 1948) 304-306.

   Hungry Gulliver, P.H. Johnson
   Thomas Wolfe, H.J. Muller

   - *O&S* (1983) 225-227.

7. *International Journal of Ethics* **60**, 2 (January 1950) 144-146.

   Nineteen Eighty-Four, George Orwell

   - *O&S* (1983) 228-231.

8. "In the Beginning the Word." *New Leader* (March 18, 1950).

   The Religious Revolt Against Reason, L.H. DeWolf

   - *O&S* (1983) 232-233.

9. "Turning the Other Cheek." *New York Review of Books* (February 21, 1974) 29-30.

   The Preachers, James Morris

   - *S:GBB* (1981) 263-273.

10. "A Passionate Realist." *New Leader* **57** (October 14, 1974) 19-20.

    The Philosophy of Karl , P.A. Schlipp, editor

    - *O&S* (1983) 295-297.
    - *TNIL* (1996) 488-490.

11. *New York Times Book Review* (August 22, 1976) 5,24.

    The Gift of Inner Healing, Ruth Carter Stapleton

    - *S:GBB* (1981) 307-313.

12. *Virginia Seminary Journal* (Winter 1976) 45-48.

    A Month of Sundays, John Updike

    - *O&S* (1983) 305-312.

13. "A Mystic Celebration." *New Leader* **60** (May 23, 1977) 23-24.

    The Tao is Silent, Raymond Smullyan

    - *O&S* (1983) 320-322.

14. George Groth [Martin Gardner], "Gardner's Game with God." *New York Review of Books* **30**, 19 (December 8, 1983) 41-43.

    The Whys of a Philosophical Scrivener, Martin Gardner

    - *W&W* (1989) 162-168.
    - *TNIL* (1996) 481-487.

15. "Giving God a Hand." *New York Review of Books* (August 13, 1987) 17-23.

    Reviews 14 books by and about Oral Roberts, Tammy Bakker, Jimmy Swaggart, and Pat Robertson.

    - *NA* (1988) 223-245.

16. "A Look at 'The Last Temptation'." *Times-News* [Hendersonville, NC] (October 30,1988) 7B.

    The Last Temptation of Christ, directed by Martin Scorsese

    - *OTWS* (1992) 127-129.

17. *The Humanist* (March/April 1991) 39-40.

    Atheism: A Philosophical Justification, Michael Martin

    - *WW&FL* (1996) 216-218.

18. "Waiting for the Last Judgement." *Washington Post Book World* (November 8, 1992) 5.

    When Time Shall Be No More, Paul Boyer

    - *WW&FL* (1996) 248-250.

19. "How He Lost It." *New York Review of Books* (May 29, 1997) 29-32.

    I Was Wrong, Jim Bakker
    Tammy: Telling It My Way, Tammy Faye Messner

    - *WJ* (2000) 149-166.

20. "Reincarnation Undressed." *Free Inquiry* (Summer 1997) 58-60.

    Reincarnation: A Critical Examination, Paul Edwards

    - *WJ* (2000) 134-141.

21. *Los Angeles Times Book Review* (April 12, 1998) 11-12.

    Nearer, My God, William F. Buckley, Jr.

    - *WJ* (2000) 332-350.

22. "Mind over Matter." *Los Angeles Times Book Review* (August 22, 1999) 4-5.

    Mary Baker Eddy, Gillian Gill
    Science and Health, Mary Baker Eddy [listed but not reviewed]
    God's Perfect Child, Caroline Fraser

    - *WJ* (2000) 179-206.

23. "The Price We Pay." *The New Criterion* (November 2008) 70-74.

    God's Problem, Bart Ehrman

    - *T&F* (2009) 212-219.

9.5 Published Correspondence

1. *Revelation* (January 1934).

   Discusses his conversion; signed "a student in the University of Chicago."

2. "The Road to Rome." *The New Republic* (December 13, 1940).

   "Pray for the conversion of Mr. Adler."

   - *The Whys of a Philosophical Scrivener* (Morrow, 1983) 407.
   - *TNIL* (1996) 549-550.

3. "Hook's Marx." *New Leader* (November, 18, 1957) 28-29.

   Objects to Sidney Hook's broad definition of pragmatism.

4. "On the Meaning of the Universe." *New York Times* (March 24, 1978) A4.

   Response to an essay by Malcolm Browne on cosmology and faith.

5. "Niebuhr & Supernaturalism." *New York Review of Books* (March 27, 1986) 52-53.

6. "Who Wants Swaggart Off the Air." *Free Inquiry* (Fall 1989) 3.

7. *Hypotheses: Neo-Aristotelian Analysis* 3 (Fall 1992) 11.

   Remarks about Adler and McKeon and how they were anticipated by H. G. Wells.

8. *Hypotheses: Neo-Aristotelian Analysis* 4 (Winter 1993) 22-23.

   Remarks about Eddy and Wilcox

9. "Gardner: Deism Confused with Christian Orthodoxy." *Times-News* [Hendersonville, NC] (August 30, 1994).

   Letter in Stephen Black's column "Mud Creek Rambler."

10. "A Friend in Prison." *New York Review of Books* (September 25, 1997) 77.

    Follow-up to his review of Bakker's book, mentioning Bakker's attorney.

11. *Los Angeles Times Book Review* (September 19, 1999) 9.

    Rejoinder about his review of Christian Science books.

    - *WJ* (2000) 204-205.

12. *Los Angeles Times Book Review* (May 3, 1998) 2.

    Rejoinder to William F. Buckley Jr

    - *WJ* (2000) 349.

13. *Free Inquiry* **20**, 2 (Spring 2000) 48.

    "Happy 20th anniversary" greeting

the new Leader

F. A. VOIGT: **WHY GERMANY WON'T FIGHT** PAGE **2**

OCTOBER 7, 1950

15 cents

Chapter 10

Political

Martin Gardner was a democratic socialist, which he explains in his *The Whys of a Philosophical Scrivener.* More work for the cause can be found in the Journalism chapter. He was a "fellow traveler" in the 1930's, but with his disillusion with the Soviet movement he naturally became critical of communist promoters.

10.1 Articles

1. "Gerald Smith in Tulsa." *The Progressive* (December 1948) 24-25.
2. "Gerald L.K. Smith Goes to Tulsa." *New Leader* **33**, 23 (June 10, 1950) 10-11.

   Much abridged.

   - *O&S* (1983) 81-85. Unedited.

3. "H.G. Wells: Premature Anti-Communist." *New Leader* **33**, 40 (October 7, 1950) 20-21.
   - *O&S* (1983) 98-102.

4. "H. G. Wells in Russia." *The Freeman* **45** (May 1995) 282-287.

   "Remembering Wells' long-forgotten 1920 book *Russia in the Shadows.*"

   - *TNIL* (1996) 140-148.

5. "Is Socialism a Dirty Word?" In *When You Were a Tadpole and I Was a Fish* (Hill and Wang, 2009) 223-226.

   From letters written to *The Norman Transcript*.

10.2   Book Reviews

1. "The Gods that Failed." *International Journal of Ethics* **60**, 4 (July 1950) 296-298.

   The Gods that Failed, Edited by R. Grossman
   The Vital Center, A. M. Schlesinger

   - *O&S* (1983) 234-237.

2. "Listing Enemies." *New Leader* **56** (July 23, 1973) 19-21.

   A Journal of the Plague Years, Stefan Kanfer

   - *O&S* (1983) 284-288.

3. "The Tangled Roots of Nazism." *Dimensions* **6**, 2 (1992) 26-28.

   The Crooked Timber of Humanity, Isaiah Berlin

   - *TNIL* (1996) 457-463.

10.3   Published Correspondence

1. "Much That is Irrelevant." *The Daily Maroon* [University of Chicago] (January 12,1937) 2.

   On the Oxford Oath.

2. "Reds are Subtle." *Tulsa World* (May 22,1938).

   Attack on General Johnson Hagood.

3. "Naive Assumption." *New Republic* (February 20, 1950) 4.

   Reply to Merle Miller about Hiss.

4. *New York Post* (January 24, 1951) M9.

   Shows Henry Wallace's ignorance is inexcusable.

5. *New Leader* (August 18, 1952) 20.

   On an article on Fabianism by Daniel Bell.

6. *New Leader* (November 18, 1957) 26.

   On Sidney Hook.

7. "Lacks Balance." *New York Times Magazine* (June 14, 1964) 22.

   Comments on an analogy of Toynbee's.

8. "Of Jimmy Carter, Church and State." *New York Times* (June 26, 1976) 22.

9. "White House Poetry." *New York Times* (October 21, 1979) 20E.

   Whimsical response to Jody Powell's comment on the Florida election.

10. "The People, Yes." *Free Inquiry* (Fall 1989) 40.

    Quotes long passages from Carl Sandburg's poem in response to the Chinese crackdown.

11. "Flora Lewis Almost Had It Right." *Times-News* [Hendersonville, NC] (June 16, 1990) 4.

    Discusses possible mixed-economies in the former USSR.

Part III

Literature and Other Topics

Martin Gardner wrote fiction and poetry. He read literature but had no interest in popular fiction. For enjoyment his personal taste ran to fantasy which explains his interest in Chesterton, Coleridge, Carroll, Dunsany and Baum.

His only major work of fiction was a theological novel. He wrote it to release the wellspring of reactions to his personal religious journey. He had other ideas for books of fiction that were never realized. (He wanted to write a book about a pair of peripatetic adventurers—like Don Quixote and Sancho Panza—that were guided by random dice.) He tried to enter the short story market for about a decade.

He always was interested in the background and/or insights found in literature. He had no interest in literary theory as found in academic treatises. Modern analysis bored him. His primary interest in literature was in understanding (not mind-reading) and then explaining it to others. He had an extensive collection of file cards on literature, and in the 1950s he tried to start a magazine that would annotate short stories.

In 1960 he began to annotate, or introduce, various editions of the works of his favorite authors. He concentrated on fantasy authors. He did not have any interest in discussing normal fiction. No one believed that the public would be interested in literature with footnotes and glosses. His work was an original approach based on his personal ability to find material that the reading public found interesting. He single-handedly rehabilitated Lewis Carroll's image, who became loved by the public and a respectable subject of scholarly investigation. The same can be said for his other subjects and the subjects of other annotated editions written by people building on Gardner's foundation. He wrote commentaries on H. G. Wells.

Gardner also annotated poetry. One-poem poets intrigued him. Modern poetry was ignored by Gardner, which he found to be lifeless and lacking an essential musicality. He made an effort, in two anthologies, to revive interest in poems that were once quite popular.

In this Part we also see Gardner's efforts for the juvenile market. Gardner never forgot his childhood influences and was grateful to those that generated his interest in games, puzzles, magic, math and a fascination with the natural world. He tried to write for the next generation. It was not a paycheck, it was "paying it forward."

There are also chapters on his journalistic efforts and a variety of miscellaneous works.

Cautiously peered Oom over the rim of the Milky Way and focused his upper eye on an atomic cloud. As it mushroomed slowly from the surface of the earth, tears dripped from the upper eye of Oom.

But a little lower than the Legnas had Oom created man. Male and female, in an image somewhat like his own, created he them. Yet ever and again had they turned the power of reason upon themselves, and the history of their race had been one of endless discord.

And Oom knew that when the atomic clouds had cleared, and bacterial plagues had spent their fury, the race of man would yet live on. After war's exhaustion would come rest and returning strength, new cities and new dreams, new loves and hates, and again the crafty planning of new wars. So might things continue until the end of space-time.

Oom wearied of man's imperfection. Sighing, lightly he touched the earth with the tip of his left big toe.

And on the earth came a mighty quaking. Lightning and thunder raged, winds blew, mountains rent asunder. Waters churned above the continents. When silence came at last, and the sea slipped back into the hollows, no living thing remained.

Throughout the cosmos other planets whirled quietly about other suns, and on each had Oom caused divers manners of souls to grow. And on each had been ceaseless bitterness and strife. One by one, gently were they touched by a toe of Oom, until the glowing suns harbored only the weaving bodies of dead worlds.

Like intricate jewelled clockwork the universe ran on. And of this clockwork Oom greatly wearied. Softly he breathed on the glittering spheres and the lights of the suns went out and a vast Darkness brooded over the deep.

In the courts of Oom many laughed at the coming of the Darkness; but others did not laugh,

OOM
by

Martin Gardner

regretting the passing of the suns. Over the justice of Oom's indignation a great quarreling arose among the Legnas, and the sound of their quarreling reached the lower ear of Oom.

Then turned Oom and fiercely looked upon the Legnas. And when they beheld his countenance they drew back in terror, their wings trailing. Gently did Oom blow his breath upon them....

And as Oom walked the empty corridors, brooding darkly on the failures of his handiwork, a great loneliness came upon him. Within him a portion of himself spoke, saying:

"Thou hast done a foolish thing. Eternity is long and Thou shalt weary of thyself."

And it angered Oom that his soul be thus divided; that in him should be this restlessness and imperfection. Even of himself Oom wearied. Raising high his middle arm into the Darkness, he made the sign of Oom.

Over infinite distances did the arm traverse. Eternity came and fled ere the sign had been completed.

Then at last to the cosmos came perfect peace, and a wandering wind of nothingness blew wistfully over the spot where Oom had been.

Figure 10.1: The entire story from *The Journal of Science Fiction.*

Fiction

He wrote only one novel, *The Flight of Peter Fromm*, that detailed his own lengthy struggle with faith. He rewrote the book several times over two decades. His second "novel" was really just the irresistible urge to contribute to the Oz literature; see the chapter on Oziana.

He seriously attempted to break into the short story market, spurred by early favorable reactions. But the short story market was shrinking even as hopeful authors increased in number. It became clear that his future was in nonfiction. Much later he anthologized these stories. Several other unsold stories remain in his files. His short stories for juvenile magazines are listed in the Juvenile chapter.

11.1 Books

1. *The Flight of Peter Fromm* (Kaufman, 1973).

   - *The Flight of Peter Fromm* (Noonday Press, 1989).
   - *The Flight of Peter Fromm* (Prometheus Books, 1994).

   New afterword.

2. *The No-Sided Professor: and Other Tales of Fantasy, Humor, Mystery, and Philosophy* (Prometheus, 1987).

   Reprints most of the short stories (as well as 2 poems) with introductions and postscripts. There are four previously unpublished pieces.

3. *Visitors from Oz* (St. Martin's Press, 1998). This citation also appears in "Oziana"

   - *Visitors from Oz* (Battered Silicon Dispatch Box, 2010).

11.2   Short Stories

1. "Thang." *Comment* [University of Chicago] (February 1936) 18.

   - "Thang." *The Folio* [Indiana University Quarterly] (April 1936) 51-52.
   - "Thang." In *The Best Science-Fiction Stories*, edited by E. F. Bleiler and T. E. Dikty (Fredrick Fell, 1949) 130-131.
   - "Thang." In *Science-Fiction Omnibus*, edited by E. F. Bleiler and T. E. Dikty (Garden City Books, 1952).

     This edition combines the 1949 and 1950 omnibuses

   - "Thang." *Allt* 11 (November 15, 1951) 46. (Danish.)
   - "Thang." In *100 Great Science-Fiction Short Short Stories*, edited by Isaac Asimov, et al. (Doubleday, 1978) 202-203.
   - *N-S P* (1987) 9-10.
   - "Thang." *Urania* 827 (March 1980) 46. (Italian.)
   - *Thang*, edited by Isaac Asimov and Martin H. Greenberg (Daw, 1983) 135-137.
   - "Thang." In *The Golden Years of Science-Fiction (Fifth Series)*, edited by Isaac Asimov and Martin H. Greenberg (Bonanza, 1983) 489-491.

     Introductory text by Asimov is the same as the Daw volume.

   - "Thang." In *The Little Book of Horrors*, edited by Sebastian Wolfe (Barricade Books, 1992) 118-119.

2. "The Dome of Many Colors." *University Review* [University of Kansas City] **11**, 2 (Winter 1944) 95-100.

   - *N-S P* (1987) 11-20.

3. "Good Dancing Sailor!" *University Review* [University of Kansas City] **12**, 3 (Spring 1946) 232-236.

   - *N-S P* (1987) 27-34.

4. "The Horse on the Escalator." *Esquire* (October 1946) 105, 235-237.

   - "The Horse on the Escalator." *Linking Ring* **58**, 5 (May 1978) 49-52.
   - *N-S P* (1987) 35-53.

5. "The No-Sided Professor." *Esquire* (January 1947) 67+.

- "Le Professor Slapenarski [French]." *Caliban* 34 (December 1949) 49-56.
- "The No-Sided Professor." *Magazine of Fantasy and Science Fiction* **2** (February 1951) 74-84.
- "The No-Sided Professor." In *Best from Fantasy and Science Fiction*, edited by Anthony Boucher and J. Francis Mc-Comas (Little, Brown, 1952) 68-81.
- "L'Homme Non Lateral [French]." *Fiction* **5**, 42 (May 1957) 99-108.
- "The No-Sided Professor." In *Fantasia Mathematica*, edited by Clifton Fadiman (Simon and Schuster, 1958) 99-109.
- "La Terrible Histoire de L'Homme Non Lateral [French]." *Planète* **12** (September-October 1963) 99-105. Illustrated by Roland Cat
- "The No-Sided Professor." In *Vintage Anthology of Science Fiction*, edited by C. Cerf (Vintage, 1966) 59-70.
- "The No-Sided Professor." In *As Tomorrow Becomes Today*, edited by C.W. Sullivan (Prentice-Hall, 1974) 127-136.
- "El Professor No-Lateral [Spanish]." *Humor and Juegos* 4 (November 1980) 68-73.
- "Profesorul fără nici o parte [Romanian]." In *Paleoaritmetică si alte probleme de logică* (1981).
- "Ein topologisches Problem [German]." In *Zweiter Almanach der Hobbit Presse*, edited by Joachim Kalka (Hobbit Presse, 1983) 136-149.
- "El Profesor No-Lateral [Spanish]." *Cacumen* **2**, 2 (March 1984) unknown.
- "A Nullaoldalú Professzor [Hungarian]." In *Piknik a Senkiflődjén* [Picnic in No Man's Land], edited by Hugó Ágoston (Kriterion Könyvkiadó, 1985) 321-335.
- "The No-Sided Professor." In *Mathenauts: Tales of Mathematical Wonder*, edited by Rudy Rucker (Arbor House, 1987) 168-178.
- *N-S P* (1987) 45-58.
- "Nul'storonii Professor." In [*Strela Vremeni*][[Arrow of Time] (Izd-vo Pravda, 1989) 67-77.
- "The No-Sided Professor." In *Fantasia Mathematica*, edited by Clifton Fadiman (Copernicus, 1997) 99-109. (Reissue of 1958 edition)

6. "The Conspicuous Turtle." *Esquire* (April 1947) 59-61, 161-164.

- *N-S P* (1987) 59-73.

7. "The Fall of Flatbush Smith." *Esquire* (September 1947) 44.

   - *N-S P* (1987) 75-79.
   - "The Fall of Flatbush Smith." In *Riffs & Choruses: A New Jazz Anthology*, edited by Andrew Clark (Continuum, 2001) 139-141.

8. "Flo's Freudian Slips." *Esquire* (October 1947) 94-95,188-189.

9. "The Lady Says 'Check'." *Esquire* (January 1948) 75.

   - "Nora Says Check." In *N-S P* (1987) 81-86.

10. "The Loves of Lady Coldpence." *Esquire* (March 1948) 40, 126, 128.

    - *N-S P* (1987) 87-99.

11. "Dr. Clodhopper's Footsies." *Esquire* (May 1948) 155-158.

    - "At the Feet of Karl Klodhopper." In *N-S P* (1987) 101-112.

12. "The Devil and the Trombone." *Record Changer* (May 1948) 10.

    - "The Devil and the Trombone." In *100 Great Science Fiction Short Short Stories*, edited by Isaac Asimov, et al (Doubleday, 1978) 27-30.
    - "The Devil and the Trombone." In *N-S P* (1987) 115-118.
    - "The Devil and the Trombone." In *Hot and Cool: Jazz Short Stories*, edited by Marcela Breton (Plume, 1990).

      Also two British editions: Penguin 1990, Bloomsbury 1991

    - "Il Diavolo e il Trombone." *Urania* [Italian] 815 (December, 1979) 6 .

13. "The Blue Birthmark." *Hence* **3**, 4 (July/August 1948) 29-31.

    - *N-S P* (1987) 119-122.

14. "The Trouble with Trombones." *Record Changer* **8**, 10 (October 1948) 10.

    - "Sibyl Sits In." In *N-S P* (1987) 123-126.

15. "Love and Tiddlywinks." *Esquire* (September 1949) 76.

    - *N-S P* (1987) 127-130.

16. "One More Martini." *Esquire* (February 1950) 83.

    - *N-S P* (1987) 131-135.

17. "Chlorophyll Phil." *Joker* **2**, 23 (1950) 80-82.

18. "That Old Man Gloom." *Esquire* (November 1950) 79.

    - "Old Man Gloom." In *N-S P* (1987) 137-141.

19. "The Horrible Horns." *London Mystery Magazine* 7 (December/January 1950-1951) 82-89.

   - *N-S P* (1987) 143-153.
   - *Mystery of an Artist: Austin Osman Spare*, edited by A. R. Naylor (I-H-O Books, 2002) 35-54.

20. "Crunchy Wunchy's First Case." *London Mystery Magazine* 8 (February/March 1951) 75-81.

   - "Meet Private Eye Oglesby." *Ellery Queen's Mystery Magazine* **44**, 6 (December 1964) 131-137.
   - "Private Eye Oglesby." In *N-S P* (1987) 155-165.

21. "Left or Right." *Esquire* (February 1951) 44-45, 111.

   - "Left or Right." In *Mathenauts: Tales of Mathematical Wonder*, edited by Rudy Rucker (Arbor House, 1987) 108-111.

22. "The Virgin From Kalamazoo." *Men Only* **47**, 186 (June 1951) 70-72.

   - *N-S P* (1987) 167-169.

23. "The Sixth Ship." *Our Navy* **46**, 7 (September 1951) 8-9.

   - *N-S P* (1987) 172-178.

24. "Oom." *Journal of Science-Fiction* **1**, 1 (Fall 1951) 13.

   - "Oom." In *100 Great Science Fiction Short Short Stories*, edited by Isaac Asimov, et al (Doubleday, 1978) 154-155.
   - *N-S P* (1987) 179-181.

25. "Merlina and the Colored Ice." *A.D.* [Anno Domini] (A Quarterly Magazine of Literature and Art) **2**, 3 (Autumn 1951) 297-303.

   - *N-S P* (1987) 183-191.

26. "Miss Medford's Moon." *Esquire* (February 1952) 34, 104, 106.

   - "Miss Medford's Moon." *Science Fiction Digest* **1**, 2 (February 1954) 39-47.

   This is a heavily revised version.

27. "The Island of Five Colors." In *Future Tense*, edited by K.E. Crossen (Greenberg, 1952) 210-228.

   Not collected in *No-Sided Professor* because it had a mathematical mistake; it confused the four-color theorem with the simpler fact that five countries cannot touch each other.

   - "The Island of Five Colors." In *Fantasia Mathematica*, edited by Clifton Fadiman (Simon and Schuster, 1958) 196-210.

28. "The Missing Walnuts." *Humpty Dumpty's Magazine* (February 1955) 76-83.

- "As Nozes Desaparecidas." *Mistério Magazine de Ellery Queen* 87 (1956) 54-56.
- "The Missing Walnuts." *Ellery Queen's Mystery Magazine* **27** (February 1956) 78-80.
- "[The Missing Walnuts]." *Revista do Globo* 87 (1957). From correspondence it also appeared in the newspaper *O Globo*
- "The Missing Walnuts." *The Cormorant's Ring* **4**, 2 (August 1997) 6-10. (A Sherlockian newsletter.)

  This was just one of nearly 100 short stories written for various children's periodicals. This one attracted more attention due to its Sherlockian allusions.

29. "The Three Cowboys." *Humpty Dumpty's Magazine* (January 1959).

- *The No-Sided Professor* (Prometheus, 1987) 193-196.

30. "Mysterious Smith." In *The No-Sided Professor* (Prometheus, 1987) 201-210.
31. "Ranklin Felano Doosevelt." In *The No-Sided Professor* (Prometheus, 1987) 211-219.
32. "The Stranger." In *The No-Sided Professor* (Prometheus, 1987) 221-224.
33. "Antimatter." In *Worlds in Small*, edited by John Robert Colombo (Cacanadadada Press, 1992) 68.

    The editor of this volume of tiny stories, found this "story" in *The Ambidextrous Universe*. In its entirety it is: "Boy meets anti-girl; they kiss; the end."

34. "From Oz to Earth." *Math Horizons* **8**, 1 (September 2000) 18-19.

    Excerpt from *Visitors from Oz* with a sidebar from *Scientific American*.

35. "Against the Odds." *The College Mathematics Journal* **32** (January 2001) 39-43.

- *UTB* (2003) 49-56.
- "Against the Odds." In *Martin Gardner in the Twenty-First Century*, edited by Michael Henle and Brian Hopkins (Math Assoc of America, 2012) 265-269.

36. "The Great Crumpled Paper Hoax." *The Ojai Orange* **54** (Summer 2007) [2-3]. ("John Wilcock's personal magazine")

- *JFH* (2008) 193-196.
- "The Great Crumpled Paper Hoax." *Philosophy Now* 153 (December/January 2022/2023).

37. "Superstrings and Thelma." *Math Horizons* **18**, 1 (September 2010) 7-9.

- "Superstrings and Thelma." In *Martin Gardner in the Twenty-First Century*, edited by Michael Henle and Brian Hopkins (Math Assoc of America, 2012) 289-291.

11.3 Published Correspondence

1. *Tabebuian* 23 (Fall 1975) 20.

Chatty letter in this rare S-F fanzine that mentions a 1976 Dell edition of *Fromm* that did not appear.

Martin Gardner

DESTINY

A leaf swayed with the breeze, then fluttering fell
In silence to the whitened slope below.
This and nothing more. Yet who can tell,
As solitary watchers of the show,
Who played the lead? Did fate or chance compel
The fall? What circumstances long ago
Then did the future of the act foretell,
And plot the spiral pathway to the snow?
So with life; tides in affairs of men
Are simply outbursts of a pondrous chain
Of countless forces brewing through the age.
The rise of empires; fruits of sword and pen
Converge and bring to focus deeds as vain
And trivial as the turning of a page.

Chapter 12

Poetry

His interest in poetry was natural to someone growing up in the 1920s. His father was a published poet and his mother would often communicate in verse. Despite his affection for Carl Sandburg he had little interest in free verse. Martin Gardner's own poetry was never much more than doggerel; he often hid behind the pseudonym Armand T. Ringer (an anagram). He was influenced by Burton Stevenson's book *Famous Single Poems* and wrote books and articles about "one-poem poets." He also published several poetry anthologies as a result of his hobby of collecting "Readers," anthologies for public speakers.

In addition to these poems, he had a regular feature in *Humpty Dumpty Magazine* of cautionary verse; see Juvenile chapter.

12.1 Poems

1. "The Swan." *School Life* [Tulsa Central High School] (1930). An acrostic on "what the hell."
2. "Destiny." *Poetry* [Tulsa Central High School] 1 (1929 - 1931) 44.
3. "A Change of Program." *School Life* [Tulsa Central High School] (January 6, 1932).
4. "Old Soldier." *School Life [Tulsa Central High School]* (February 24, 1932) 2.

5. "[Various Poems]." In *Essays in Prose and Verse*, edited by [Lev-Ellen Gilliam] ([Tulsa Central High School], 1932).

The Old Soldier, p. 25
Life, p. 26
Red, p. 27
An Ethraldrian Gazes at the Earth, p. 27
The Pause, p. 27
A Change of Program, p. 29
If Any Dare, p. 29

6. [Martin Gardner], "To A Dead Body Washed Ashore." *Comment (University of Chicago)* **3**, 2 (Winter 1935) 7.
7. "Chicago." *Man!* (May 1936) 7.
8. [Martin] Gardner and [Marian] Wagner, *Scrubwoman's Lament* (Blackfriars (University of Chicago), May 1936).

This citation is for the lyrics to a song written for the 32nd annual Blackfriars' musical "Fascist and Furious".

9. [Martin] Gardner and [Marian] Wagner, *Brigadier General McGlurk* (Blackfriars (University of Chicago), May 1936).

From the same show as above.

10. "Moon Probes." *Stranger Than Fact* (March 1964) 13.

This appeared in a literary S-F fanzine, edited by Jim Harkness.

11. Nitram Rendrag [Martin Gardner], "Casey's Son." In *The Annotated Casey at the Bat* (Potter, 1967) 102-104.
    - "The Son of Mighty Casey: A Tale of a Terrible Downfall." In *N-S P* (1987) 197-199.

12. "Fractal Song: Towards a Brownian Poem." *Star Line: Newsletter of the Science Fiction Poetry Association* **3**, 2 (March/April 1980) 22.

Apparently a "found poem" from an unknown bit of prose; it is:

"When we analyze the dynamic world, made up
of quantities constantly changing in time,
we find a wealth of fractal-like fluctuations
that have $1/f$ spectral densities."

13. "So Long Old Gal." In *The No-Sided Professor* (Prometheus, 1987) 21-25.

   - *JFH* (2008) 197-201.
   - "Martin Gardner, Yeoman/2c." In *G4G11 Exchange Book: Volume 2* (Gathering4Gardner, 2017) 43-48.

   This is a version of the poem annotated by Dana Richards

14. Armand T. Ringer [Martin Gardner], "Santa Changes His Mind." In *The Annotated Night Before Christmas* (Summit, 1991) 85-86.
15. Nitram Rendrag [Martin Gardner], "Mighty Yeltsin." In *The Annotated Casey at the Bat (Third, Revised Edition)* (Dover, 1995) 230-231.
16. Armand T. Ringer [Martin Gardner], "Mathematical Realism and Its Discontents." *Los Angeles Times Book Review* (October 12, 1997) 8.

   Limerick on T.O.E. as an epigram to book review

   - *WJ* (2000) 227.

17. Armand T. Ringer [Martin Gardner], "As a photon flew close to slit 2 ...." *Skeptical Inquirer* **24**, 3 (May/June 2000) 13.

   A limerick embedded in an otherwise serious Gardner column on de Broglie and Bohm. Also on the same page is an original Eddie Guest parody about von Neuman.

18. Armand T. Ringer [Martin Gardner], "Little Red Riding Hood." *Skeptical Inquirer* **24**, 5 (September/October 2000) 14.

   Epigram for a column of Bettelheim's analysis.

19. Armand T. Ringer [Martin Gardner], "Six Poems." *Hypotheses* 31-32 (Fall 1999/Winter 2000) 8.

   Poems that were to appear in the forthcoming *Poetic Parodies*; they are explained in a "Correspondence" on p. 12.

20. "The Minnesota Wrestler." *Free Inquiry* **20**, 2 (Spring 2000) 7.

   Poem from the forthcoming *Poetic Parodies*, appeared under his own name.

21. Armand T. Ringer [Martin Gardner], "[Various Poems]." In *Martin Gardner's Favorite Poetic Parodies*, edited by Martin Gardner (Prometheus, 2001).

Memory, p. 17
The Closet Has a Hundred Ties, p. 23
The Web Links, p. 25
The Ancient Shortstop, p. 31
The Lady Explains, p. 34
A House Far Away from the Road, p. 41-42
The Road I Took, p. 44
Warm or Cold, p. 47
When It Can't Be Done, p. 50
Fleas, p. 65-66
The Minnesota Wrestler, p. 96-97
Road Fog, p. 161
Further Thoughts, p. 175-177
Twinkle, Little Star, p. 180
To Shirley Temple, p. 180
Leaving the Bar, p. 182-183
The Bold Spoken Lass Who Hung Out at the Well, p. 226-227
Mary's Naughty Cat, p. 229
The Purple Skunk, p. 239

22. "Three Parodies of Famous Poems." In *Are Universes Thicker than Blackberries?* (Norton, 2003) 158-160.

Previously unpublished parodies:
"What's Done is Done," (apologies to Longfellow)
"Jackson Pollack," (apologies to Kipling)
"The Holdup Man," (apologies to Noyes)

23. Armand T. Ringer [Martin Gardner], "Ode to Apricots." *Word Ways* **40**, 1 (February 2007) 38-39.

This is one of seven pieces of material (pp. 36-39) that Gardner contributed to David Morice's "Kickshaws" column. Most were from other people but they also included two new quatrains by Gardner: "Tweedle D.D" and "Tom's Height."

12.2   Published Correspondence

1. *Hypotheses: Neo-Aristotelian Analysis* 6 (Summer 1993) 18.

Chatty letter about the Dover books of poems.

Chapter 13

Literature

The idea that literature can be annotated for the general public seems to be largely original with Martin Gardner. While there might be minor precursors his vision was original. It was immediately successful and has spawned at least 185 other "Annotated" texts all based on his vision of bridging between the experts and the public. *The Annotated Alice* (1960) was first and is discussed in the Carrolliana chapter. His writings on L. Frank Baum are covered in the Oziana chapter.

He was interested in poetry, and even advertised himself as a poet for hire. He was dismissive of modern poetry, even though he greatly impressed by the free verse of Carl Sandburg. Late in life he collected "readers" from used bookstores, books with verse meant to be read aloud socially. He used these to assemble books of "popular" poems.

His early interest in literature is seen in the column that he wrote for the University of Chicago student newspaper.

13.1 Books

1. *The Fantastic Fiction of Gilbert Chesterton* (The Battered Silicon Dispatch Box, 2008).

   With a foreword by John Peterson, an afterword by Pasquale Accardo, and four appendices of illustrations. Contains five new chapters and reprints of other articles and book introductions.

13.2   Annotated Editions

1. Samuel Taylor Coleridge, *The Annotated Ancient Mariner* (Potter, 1965).

   - Samuel Taylor Coleridge, *The Annotated Ancient Mariner* (Bramhall House, 1965).
   - Samuel Taylor Coleridge, *The Annotated Ancient Mariner* (Meridian Books / World Publishing Co, 1967).
   - Samuel Taylor Coleridge, *The Annotated Ancient Mariner* (Meridian Books / New American Library, 1974).
   - *W&W* (1989) 3-24.

     Reprints just the Afterword.

   - *TNIL* (1996) 297-317.

     Reprints just the Introduction.

   - Samuel Taylor Coleridge, *The Annotated Ancient Mariner* [Second Edition] (Prometheus Books, 2003).

     New preface, six new footnotes and some corrections.

2. Ernest Lawrence Thayer, *The Annotated Casey at the Bat: A Collection of Ballads about the Mighty Casey* (Potter, 1967).

   - Ernest Lawrence Thayer, *The Annotated Casey at the Bat: A Collection of Ballads about the Mighty Casey* (Bramhall House, 1967).
   - *The Annotated Casey at the Bat: A Collection of Ballads about the Mighty Casey (Second Edition)* (University of Chicago Press, 1984).

     Appendix II added.

   - *The Annotated Casey at the Bat: A Collection of Ballads about the Mighty Casey (Third, Revised Edition)* (Dover, 1995).

     Appendix III added.

   - *The Annotated Casey at the Bat: A Collection of Ballads about the Mighty Casey (Fourth, Revised Edition)* (The Battered Silicon Dispatch Box, 2010).

     "Revised and corrected," with an expanded appendix

3. G. K. Chesterton, *The Annotated Innocence of Father Brown* (Oxford University Press, 1987).

- G. K. Chesterton, *The Annotated Innocence of Father Brown* (Dover, 1987).

  Revised with a new introduction. First U.S. edition.

- *FFGC* (2008) 80-96.

4. *The Annotated Night Before Christmas: A Collection of Sequels, Parodies, and Imitations of Clement Moore's Immortal Ballad about Santa Claus* (Summit, 1991).

   - *The Annotated Night Before Christmas: A Collection of Sequels, Parodies, and Imitations of Clement Moore's Immortal Ballad about Santa Claus* (Prometheus, 2005).
   - *JFH* (2008) 173-191.

     Introduction only.

5. *Best Remembered Poems* (Dover, 1992).
6. *Famous Poems From Bygone Days* (Dover, 1995).
7. H. G. Wells, *"The Country of the Blind" and Other Science-Fiction Stories* (Dover, 1997).

   - *WJ* (2000) 238-250. (The second third of chapter 23.)

8. G. K. Chesterton, *The Annotated Thursday* (Ignatius Press, 1999).

   - "The Man Who Was Thursday: Revisiting Chesterton's Masterpiece." *Books & Culture: A Christian Review* **6**, 3 (May/June 2000) 30-33.

     Excerpt

   - "Chesterton's *The Man Who Was Thursday*." In *UTB* (2003) 125-134.

     Reprints introduction only.

   - G. K. Chesterton, *The Man Who Was Thursday: A Nightmare* (Ignatius Press, [2004]).

     Undated second printing, with new title, new cover (top hat and cane), and corrections; no mention of second printing but see, for example, page 184: "After writing the above note for this book's first printing, I was informed ..."

   - *FFGC* (2008) 39-65.

9. *Martin Gardner's Favorite Poetic Parodies*, edited by Martin Gardner (Prometheus, 2001).

   Contains numerous poems by Martin Gardner under the pseudonym Armand T. Ringer. (See Poetry chapter for details.)

13.3   Columns

1. "Local Literati." *The Daily Maroon* [University of Chicago].

   An irregular column about books at Chicago

   12/14/34—p. 3—Thornton Wilder, Gertrude Stein, Billy Sunday [Ben Reitman]
   1/10/35—p. 3—C. S. Boucher, Thornton Wilder, and chat
   2/6/35—p. 4—A. Eustace Haydon, Martin Freeman, Janet Fairbank, Jocque Campeau, and chat
   3/7/35—p. 3—Alexander Woolcott, Archibald MacLeish, Vincent Sheean and Vardis Fisher
   10/3/35—p. 4—Edgar J. Goodspeed and James Weber Linn
   10/11/35—p. 3—James T. Farrell, Christopher Morley, coach Shaughnessy, and John Dos Passos
   10/17/35—p. 3—Vincent Sheean and chat about Thornton Wilder and Ben Reitman
   10/24/35—p. 3—Wilhelm Pauck [article was cut short]
   11/1/35—p. 3—Norman Thomas, Preston Bradley, Ernest Freemont Tittle and chat
   11/14/35—p. 3—Arthur Holly Compton
   12/4/35—p. 3—Mahanamabrata Brahmacheri and Sterling North
   1/9/36—p. 3—Gertrude Stein
   1/23/36—p. 2—Carl Sandburg
   4/7/36—p. 3—Robert M. Hutchins
   4/29/36—p. 3—Logan Clendening, Charlotte Wilder, H. Nelson Wieman, and chat

13.4   Book Introductions

1. Ray Bradbury, *The Martian Chronicles* (The Limited Editions Club [Doubleday], 1974) v-xiii.
   - *W&W* (1989) 37-43.

2. Ernest Lawrence Thayer, *Casey at the Bat* (Dover, 1977). (2012 edition with new cover)

   Has the *Annotated Casey* introduction. Illustrated by Jim Hall.

3. Lord Dunsany, *A Dreamer's Tales* (Owlswick Press, 1979) vii-xii.
   - *W&W* (1989) 44-47.

4. G. K. Chesterton, *The Club of Queer Trades* (Dover, 1987) iii-vii.
   - *FFGC* (2008) 33-38.
5. G. K. Chesterton, *Four Faultless Felons* (Dover, 1989) vii-xvi.
   - *FFGC* (2008) 159-168.
6. G. K. Chesterton, *The Paradoxes of Mr. Pond* (Dover, 1990) v-xi.
   - *FFGC* (2008) 169-175.
7. G. K. Chesterton, *The Napoleon of Notting Hill* (Dover, 1991) vii-xx.
   - *WJ* (2000) 315-331.
   - *FFGC* (2008) 16-32.
8. Frank Jacobs, *Pitiless Parodies and Other Outrageous Verse* (Dover, 1994) vii-viii.
9. G. K. Chesterton, *The Ball and the Cross* (Dover, 1995) iii-xii.
   - *WJ* (2000) 299-314.
   - *FFGC* (2008) 66-79.
10. H. G. Wells, *The Conquest of Time* (Prometheus, 1995) 11-14.
    - *WJ* (2000) 235-238.

    The first third of chapter 23.
11. Everett F. Bleiler, *Science Fiction: The Gernsback Years* (Kent State Univ Pr, 1998) xi.
12. G. K. Chesterton, *The Coloured Lands* (Dover, 2009) 233-237.
    - *T&F* (2009) 207-211.

    Poets and Lunatics: Episodes in the Life of Gabriel Gale by G. K. Chesterton (Dover 2010) was advertised as having a Martin Gardner introduction, but he died before it could be included.

13.5 Articles

1. "Will Cuppy: An Interview." *University of Chicago Magazine* (June 1941 (Part I)) 15.
2. "Thornton Wilder and the Problem of Providence." *University Review (University of Kansas City)* **10**, 3 (Spring 1944) 213-223.
3. "Art, Propaganda and Propaganda Art." *University Review (University of Kansas City)* **10**, 3 (Spring 1944) 213-223.
   - *O&S* (1983) 29-46.
4. "Humorous Science Fiction." *Writer* **62** (April 1949) 148-151.

5. "Sidney Sime of Worplesdon." *Arkham Sampler* **2**, 4 (Autumn 1949) 81-87.

   - *O&S* (1983) 47-56.
   - "Sidney Sime of Worplesdon." In *Arkham Sampler: Vol 2*, edited by August Derleth (2010) 81-87.

   A facsimile of the 1949 issues.

6. "The Golden *Galaxy*." *Journal of Science Fiction* **1**, 1 (Fall 1951) 4-6, 15.

7. "When You Were a Tadpole and I Was Fish." *Antioch Review* **22** (Fall 1962) 332-340.

   - *O&S* (1983) 147-156.
   - *T&F* (2009) 164-177.

8. "The Harvard Man Who Put the Ease in Casey's Manner." *Sports Illustrated* **22** (May 24, 1965 [Eastern Edition]) M3-4.

   Also in June 28, 1965 [Western Edition] and August 2, 1965 [Midwest Edition]

   - "The Harvard Man Who Put the Ease in Casey's Manner." In *Edge of Awareness*, edited by Ned Hoopes (Dell, 1966) 112-120.

     Also Delacorte/Laurel-Leaf (1970).
   - "Casey at the Bat." *American Heritage* **18** (October 1967) 64-68.
   - "Everybody Knows Poetic 'Casey at the Bat'." *Staten Island Advance* (April 7, 1968).

     "The Best from *American Heritage*." The next is the same.
   - "Everybody Knows Poetic 'Casey at the Bat'." *Long Island Press* (April 7, 1968).
   - "'...there is no joy in Mudville ...'." *University of Chicago Magazine* **77** (Summer 1985) 28-31.

     Excerpted from *The Annotated Casey*
   - *W&W* (1989) 25-36.

     The expanded version used in *The Annotated Casey*.
   - "The Harvard Man Who Put the Ease in Casey's Manner." In *Discovering Language*, edited by William Vesterman (Allyn and Bacon, 1992) 391-398. With study questions, pp. 398-399.

9. Martin Gardner and J.A. Lindon, "Pied Poetry." *Word Ways* **6**, 2 (May 1973) 98-100.

   • "Pied Poetry." *Word Ways* **43**, 3 (August 2010) 214-215.

10. "A 'Contradiction' in The Man Who Was Thursday?" *Chesterton Review* **1**, 2 (Spring/ Summer 1975) 124-125.

11. "Commentary (On Asimov's *Foundation* Trilogy)." *Isaac Asimov's Science Fiction Magazine* (December 1982) 75. (Just one of many commentaries on the same subject.)

12. "The Puzzles in *Ulysses*." *Semiotica* **57** (1985) 317-330.

   • *W&W* (1989) 107-121.
   • *TNIL* (1996) 354-367.

13. "H. G. Wells." In *Supernatural Fiction Writers*, edited by E. F. Bleiler (Charles Scribner's Sons, 1985) 397-402.

   • *W&W* (1989) 122-130.

14. "G. K. Chesterton." In *Supernatural Fiction Writers*, edited by E. F. Bleiler (Charles Scribner's Sons, 1985) 411-414.

   • *W&W* (1989) 131-137.

15. "Lord Dunsany." In *Supernatural Fiction Writers*, edited by E. F. Bleiler (Charles Scribner's Sons, 1985) 471-478.

   • *W&W* (1989) 138-150.

16. "John Martin's Book: An Almost Forgotten Children's Magazine." In *Children's Literature* **18**, edited by F. Butler et al. (Yale University Press, 1990) 145-159.

   This was shortened for the introduction to *Peter Puzzlemaker*, see Juvenile chapter.

   • *WJ* (2000) 77-96.

17. "Chesterton's Pump Street Hoax." *Midwest Chesterton News* **3**, 3 (May 10, 1991) 4.

18. *Washington Post Book World* (May 12, 1996) 14.

   Rejoinder to letters about Sagan's book.

   • *WJ* (2000) 330-331.

19. "Levels of Allegory in *The Ball and the Cross*." *Chesterton Review* **18**, 1 (February 1992) 37-47.

20. "I, the Jury." *Washington Post Book World* (December 5, 1993) 5.

    A one paragraph response to a request for a nomination for the Nobel Prize in Literature; Gardner nominated John Updike.

21. "The Gifts They Love to Give." *Washington Post Book World* (December 8, 1996) 4.

    A collection of book suggestions from various authors; Gardner suggested *The Man Who Was Thursday*, by G. K. Chesterton.

22. "My Favorite Mystery." *Gilbert!* **2**, 3 (December 1998) 22.

23. "More Pictures for Thursday." *Gilbert!* (January/February 2000) 11.

24. "Looking Backward at Edward Bellamy's Utopia." *The New Criterion* (September 2000) 19-25.

    - *WJ* (2000) 263-276.
    - "Looking Backward at Edward Bellamy's Utopia." In *Counterpoints*, edited by Roger Kimball and Hilton Kramer (Ivan R. Dee, 2007) 130-139.

25. "Hugo Gernsback." In *From the Wandering Jew to William F. Buckley Jr.* (Prometheus, 2000) 104-122.

    First appearance in print.

26. "Edgar Wallace and *The Green Archer*." In *Are Universes Thicker than Blackberries?* (Norton, 2003) 161-172.

    First appearance in print. Written as an introduction for an unrealized reprint edition.

27. "Afterword to *The Green Archer*." In *Are Universes Thicker than Blackberries?* (Norton, 2003) 173-177.

    First appearance in print. Written for an unrealized reprint edition.

28. "Naughts & Crescents." *The New Criterion* (March 2006) 34-36.
    - *JFH* (2008) 155-159.
    - "Chesterton's *Flying Inn*." In *FFGC* (2008) 110-117.

29. "The Fantastic Fiction of Gilbert Chesterton." *Gilbert* **11**, 4 (January/February 2008) 18-21.
    - "Manalive." In *JFH* (2008) 161-171.
    - "Chesterton's *Manalive*." In *FFGC* (2008) 101-111.

30. "Tales of the Long Bow." In *The Fantastic Fiction of Gilbert Chesterton* (Battered Silicon Dispatch Box, 2008) 101-111.

First appearance

- *T&F* (2009) 155-163.

13.6   Book Reviews

1. *New York Times Book Review* (January 18, 1976) 8, 12-13.

   Cobwebs to Catch Flies, Joyce Whalley
   Early Childrens Books and Their Illustration, G. Gottlieb and J.H. Plumb

   - *O&S* (1983) 292-294.

2. "Children's Books Aren't Just Kids Stuff." *Washington Post Book World* (October 23, 1978) 4.

   Fairy Tales and After, Roger Sale

   - *O&S* (1983) 326-327.

3. *Baker Street Miscellanea* 40 (Winter 1984) 47-49.

   Some Notes on a Meeting at Chisham, Robert Bayer

   - *W&W* (1989) 191-193.
   - "Did Sherlock Holmes Meet Father Brown." In *FFGC* (2008) 97-100.

4. "Through the Looking-Glass." *Dimensions* **4**, 3 (1989) 35-37.

   W: Or the Memory of Childhood, Georges Perec

   - *TNIL* (1996) 348-353.

5. "William Shakespeare: By Divers Hands." *Washington Post Book World* (January 19, 1992) 5.

   Shakespeare's Lives: New Edition, S. Schoenbaum

   - *TNIL* (1996) 368-374.

13.7   Published Correspondence

1. "Wintry Smile." *New York Times* (Nov 24, 1957) 301.

   In "Queries and Notes"—identifies the stanza by Alfred Noyes that "B.R." had quoted on Nov 3.

2. *New York Times Book Review* (April 30, 1961) 36.

   Corrects an assertion about Chesterton.

3. "'Casey's' Origin." *New York Times* (October 22, 1966).

   Says Thayer did not intend Mudville to be Baltimore.

4. *Chesterton Review* **11** (May 1985) 248-250.

   Sixteen queries about *The Innocence of Father Brown.*

5. "The New Chesterton Biography." *Chesterton Review* **11** (May 1985) 258.

   Corrects Mills's assertion about *Club of Queer Trades* illustrations

6. "Extraordinary Holmes!" *New York Times Book Review* (February 1, 1987) 33.

   Corrects an assertion about Chesterton.

7. *Baker Street Miscellanea* 51 (Autumn 1987) 38-39.

   About Holmes and Chesterton

8. *Chesterton Review* **13** (November 1987) 557-558.

   A reply to part I of Owen Edwards' review.

9. *Chesterton Review* **14** (May 1988) 352-353.

   A reply to part II of Owen Edwards' review.

10. "Readers' Choice." *News and Observer* [Raleigh, NC] (December 25, 1988) 5D.

    Just an excerpt from F. Scott Fitzgerald's story "Absolution" suggested by Gardner. "The book pages periodically reprint passages readers think exemplary English prose."

11. *Midwest Chesterton News* **3**, 5 (1991) 10.

    Asks a question about *The Ball and the Cross*

12. *Midwest Chesterton News* **3**, 4 (January 10, 1991) 16.

   Chatty letter; says he would like to work on *Manalive*

13. *Washington Post Book World* (Mar 8, 1992) 14.

   Rejoinder on Shakespeare, to Thomas H. Taylor.

   - *TNIL* (1996) 372-373.

14. *Washington Post Book World* (April 12, 1992) 14.

   Another rejoinder on Shakespeare, to Charlton Ogburn.

15. "A Caricature of Chesterton." *The Chesterton Review* **19**, 1 (February 1993) 135.

   A connection between Wells and Chesterton

16. *Hypotheses: Neo-Aristotelian Analysis* 8 (Winter 1994) 32.

   A request for parodies of poems; apparently also in issue no. 3, Fall 1992.

17. *Hypotheses: Neo-Aristotelian Analysis* 9 (Spring 1994) 28.

   About old University of Chicago personalities and literature.

18. *Hypotheses: Neo-Aristotelian Analysis* 19 (Fall 1996) 41.

   Remarks about Chesterton

19. *Gilbert Magazine* **19** (April/May 2006) 5.

   Remarks about Biblical humor.

20. *Gilbert Magazine* **19** (July/August 2006) 5.

   Remarks about the word Heckmondwyke.

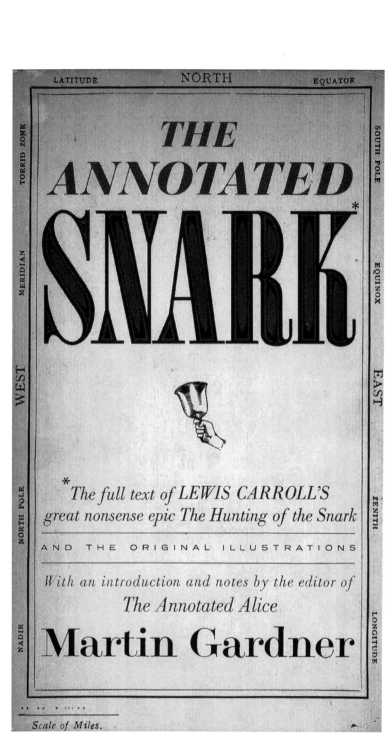

LATITUDE   NORTH   EQUATOR

TORRID ZONE   MERIDIAN   WEST   NORTH POLE   NADIR

SOUTH POLE   EQUINOX   EAST   ZENITH   LONGITUDE

THE
ANNOTATED
SNARK *

* The full text of LEWIS CARROLL'S
great nonsense epic The Hunting of the Snark

AND THE ORIGINAL ILLUSTRATIONS

With an introduction and notes by the editor of
The Annotated Alice

Martin Gardner

Scale of Miles.

Chapter 14

Carrolliana

His vision of bridging between the experts and the public began with *The Annotated Alice* (1960) and was always his best-selling book. He co-founded the Lewis Carroll Society of North America, in which he remained involved. A festschrift that celebrates his influence, *A Bouquet for the Gardener* (sic), is discussed in Section A.6.

14.1 Annotated Editions

1. Lewis Carroll, *The Annotated Alice* (Potter, 1960).

   - Lewis Carroll, *The Annotated Alice* (Bookcraftsmen Associates, 1960).
   - Lewis Carroll, *The Annotated Alice* (World Publishing / Forum Books, 1963).
   - Lewis Carroll, *The Annotated Alice* (Bramhall House, 1965).
   - Lewis Carroll, *The Annotated Alice* (New American Library / Meridian / Times Mirror, 1974).
   - Lewis Carroll, *The Annotated Alice* (Wings / Random House, 1993).

     A blue and tan dustjacket. The second printing ("updated" on title page) has a "Preface to 1995 Edition," which lists the 17 pages which contain changes to some annotations.

   - Lewis Carroll, *The Annotated Alice* (Wings / Random House, 1998).

     White dustjacket, slightly reduced pages

- *TNIL* (1996) 318-327.

  Reprints just the introduction.

2. Lewis Carroll, *The Annotated Snark* (Bramhall, 1962).

   - *The Annotated Snark* (Penguin, 1967).

     Contains a new preface.

   - "The Annotated Snark." In *The Hunting of the Snark*, edited by James Tanis and John Dooley (Kaufman, 1981) 3-79.

     Contains the complete text of his book of the same name with numerous updates for this Centennial edition of the *Snark*.

   - *The Annotated Hunting of the Snark* (Norton, 2006).

     Contains revisions, a new preface, and new appendices. Introduction by Adam Gopnik

3. Lewis Carroll, *The Wasp in the Wig* (Potter, 1977).

   - Lewis Carroll, *The Wasp in the Wig* (The Lewis Carroll Society of North America, 1977).

     Hardcover with red boards and gold rules and title on spine

   - Lewis Carroll, *The Wasp in the Wig: Carroll Studies No. 2* (The Lewis Carroll Society of North America, 1977).

     Second in an "official series of chapbooks"; red embossed wrappers with titles, etc in black

   - *More Annotated Alice* (Random House, 1990) 325-357.
   - *Annotated Alice: The Definitive Edition* (Norton, 2000) 275-298.
   - *Annotated Alice: The 150th Anniversary Deluxe Edition* (Norton, 2015) 321-345.

4. *More Annotated Alice: Alice's Adventures in Wonderland and Through the Looking Glass and What She Found There* (Random House, 1990).

   All new material plus the text of the annotated *The Wasp in the Wig*. Decorative borders by Mikhail Ivenitsky. Illustrated by Peter Newell

5. Lewis Carroll, *The Complete Annotated Alice* (Voyager Expanded Books, 1991).

   This is a machine-readable book, and may be one of the first three books released in this format. Contains the material from *The Annotated Alice*, *The Wasp in the Wig*, and *More Annotated Alice*.

6. Lewis Carroll, *Phantasmagoria* (Prometheus, 1998).

   - *JFH* (2008) 281-286.

7. *Annotated Alice: The Definitive Edition* (Norton, 2000).

   "This edition combines the notes of Gardner's 1960, *The Annotated Alice* with his 1990 update *More Annotated Alice*, as well as additional discoveries and updates." In 2006 an audio version was released by Recording for the Blind and Dyslexic, Princeton NJ.

8. *Annotated Alice: The 150th Anniversary Deluxe Edition* (Norton, 2015).

   This updated edition is edited by Mark Burstein, contains Gardner's additional discoveries and update after the *Definitive Edition* plus additional notes by the Editor. Profusely illustrated.

14.2   Book Introductions

1. Lewis Carroll, *Alice's Adventure Underground* (Dover, 1965) v-xi.
   - Lewis Carroll, *Alice's Adventure Underground* (McGraw-Hill, 1966).
   - *JFH* (2008) 293-300.

2. Lewis Carroll, *The Nursery Alice* (Dover, 1966) v-xi.
   - Lewis Carroll, *The Nursery Alice* (McGraw-Hill, 1966).
   - *JFH* (2008) 287-292.

3. Lewis Carroll, *The Hunting of the Snark* (London: Catalpa Press / Ian Hodgkins and Co Ltd, 1974). Limited to 250 copies.

4. Lewis Carroll, *Sylvie and Bruno* (Dover, 1988) v-xix.

   In 2003 an audio version was released by Recording for the Blind and Dyslexic, Princeton NJ.

   - *JFH* (2008) 265-280.

5. Lewis Carroll, *Alice's Adventures in Wonderland* (Signet Classic, 2000) v-xii.

In 2003 an audio version was released by Recording for the Blind and Dyslexic, Princeton NJ.

- *JFH* (2008) 301-307.
- Lewis Carroll, *Alice's Adventures in Wonderland* (Signet Classic, 2012) v-xii.

Reissued with a new afterword by Jeffrey Meyers

14.3   Articles

1. "Some Annotations on Jabberwocky...." *The Best* (Fall 1964) 78-86.

Taken from the *Annotated Alice*. This was short-lived periodical and is scarce.

2. "Speak Roughly." In *Lewis Carroll Observed*, edited by E. Guiliano (Potter, 1976) 19-30.

- *O&S* (1983) 173-187.

3. "Which Way to Go." *Pediatrics* **62**, 2 (August 1978) 227.

Excerpt from *Annotated Alice* used as filler.

4. "Who Wrote 'Speak Gently'? An Up Date." *Jabberwocky* (Summer 1985) 63-64.

5. "From *The Annotated Alice*." *Parabola* **11**, 2 (May 1986) 38-41.

Reprint of two long notes from *Alice*, in a special issue on "Mirrors."

6. "Well, You Know ...." *Knight Letter* 65 (Winter 2000) 7.

Written for a planned but unrealized reprint edition.

- "Well, You Know." In *A Bouquet for the Gardener: Martin Gardner Remembered*, edited by Mark Burstein (The Lewis Carroll Society of North America, 2011) 132-133.

7. "A Gardner's Bouquet: New Annotations." *Knight Letter* **11** (Summer 2005) 1-6.

See next citation.

8. "A Gardner's Nosegay: Further Annotations." *Knight Letter* **11** (Spring 2005) 1-3.

- "The Final Annotations." In *A Bouquet for the Gardener: Martin Gardner Remembered* (The Lewis Carroll Society of North America, 2011) 134-148.

    Combines the two previous articles.

9. "The Mad Gardner's Tale." *Knight Letter* 104 (Spring 2020) 35.

    Reprints a figurative poem from the back of the 1965 Penguin edition of *The Annotated Alice* that is uncredited (and may have been written by a copy editor).

14.4   Book Reviews

1. *Semiotica* **5**, 1 (1972) 89-92.

    Language and Lewis Carroll, R.D. Sutherland

    - *O&S* (1983) 280-283.

2. "Magical Mystery Tour." *Washington Post Book World* (October 14, 1973) 4.

    The Magic of Lewis Carroll, John Fisher

    - *O&S* (1983) 289-291.

3. "Through the Looking-Glass." *Nature* **379** (January 11, 1996) 127-128.

    Lewis Carroll: A Biography, Morton N. Cohen

    - *WJ* (2000) 167-170.

4. "Breathless." *Los Angles Times Book Review* (Dec. 6, 1998) 22.

    Reflections in a Looking Glass, Morton N. Cohen

    - *WJ* (2000) 171-175.

14.5   Published Correspondence

1. "Letters to the Editor." *Jabberwocky* **21**, 2 (Spring 1992) 55.

    Note on Carroll's variation of Nine Men's Morris.

2. "Leaves from the Deanery Garden." *Knight Letter* 65 (Winter 2000) 13.

   Remark on the drawing of the Jack of Hearts

3. "Anne Thackeray's *From an Island*." *The Carrollian: The Lewis Carroll Journal* 13 (Spring 2004) 57.

   Response to an essay by Karoline Leach

4. "Letter." *Bandersnatch* 128 (July 2005) 9.

   Remarks about Baum and Carroll.

5. "Replies to Melanie Bayley." *Word Ways* **43**, 3 (August 2010) 173-174.

   A letter to Jeremiah Farrell about articles by Bayley, reproduced as a photocopy (along with another letter by Solomon W. Golomb).

Chapter **15**

Oziana

The material in this chapter could have been included in the Literature chapter, but somehow it seemed better to separate it as an example of his ability to focus and build a community. He was a founding member of the International Wizard of Oz Club in 1957. Martin Gardner had a deep affection for Baum's books and he recalled reading them along with his parents. He referred to them as America's only "fairy tales." He resisted requests to annotate them, preferring to have his friend Michael Patrick Hearn do that.

15.1 Books

1. *Visitors from Oz* (St. Martin's Press, 1998). (This citation also appears in "Fiction".)

   - *Visitors from Oz* (Battered Silicon Dispatch Box, 2010).

15.2 Book Introductions

1. [L. F. Baum], *The Wizard of Oz and Who He Was* (Michigan State University Press, 1957) 19-45.

   Gardner and R. B. Nye co-edited this edition; only their names, not Baum's, appear on the title page. Gardner's biographical introduction combines the text of the two "The Royal Historian of Oz" articles. Further, Gardner provided light annotations.

- L. F. Baum, *The Wizard of Oz and Who He Was* (Michigan State University Press, 1994).

  With a new introduction by Maurice Hungiville.

- L. F. Baum, *The Wizard of Oz and Who He Was* (Literary Licensing, 2011).

2. L. F. Baum, *Wonderful Wizard of Oz* (Dover, 1960) 1-5.

   - L. F. Baum, *Wonderful Wizard of Oz* (Dover, 1960).

     Included in a 1961 boxed set containing a reading by Marvin Miller on four LPs published in The Living Literature Series. It does not state that it contains this edition, but known copies have it and it fits perfectly in the space provided.

   - L. F. Baum, *Wonderful Wizard of Oz and The Marvelous Land of Oz* (Dover, 1966).

     Issued as a two-volume hardcover boxed set, but there is an introduction only for *Wizard* as the introduction to *Land* was not published until 1969.

   - *T&F* (2009) 141-144.

     Reprints the Introduction.

3. L. F. Baum, *The Surprising Adventures of the Magical Monarch of Mo and his People* (Dover, 1968) v-xi.

   Just *The Magical Monarch of Mo* on the cover.

   - *JFH* (2008) 235-241.

4. L. F. Baum, *Marvelous Land of Oz* (Dover, 1969) v-xv.

   - *WJ* (2000) 289-298.

5. L. F. Baum, *Queen Zixi of Ix* (Dover, 1971) v-xii.

   - *JFH* (2008) 205-212.

6. L. F. Baum, *John Dough and the Cherub* (Dover, 1974) v-xxi.

   This was originally an article; see below.

7. L. F. Baum, *The Life and Adventures of Santa Claus* (Dover, 1975) ix-xxi.

   - *T&F* (2009) 145-153.

     Reprints the Introduction.

8. L. F. Baum, *American Fairy Tales* (Dover, 1978) v-xiii.

   - *JFH* (2008) 243-249.

9. *L. Frank Baum and Related Oziana* [Catalog of the Auction of the Justin Schiller Collection, Sale 1118] (New York: Swann Galleries, November 2, 1978) [vii-viii].

10. L. Frank Baum, *The Scarecrow of Oz* (The International Wizard of Oz Club, 1991) 289-302.

    - *WJ* (2000) 277-288.

11. Michael P. Hearn, *Annotated Wizard of Oz: Centennial Edition* (Norton, 2000) xi-xii. (Annotated by Michael Patrick Hearn.)

12. L. Frank Baum, *Stories of Oz* (Dover, 2011) vii-xii.

15.3   Articles

1. "The Royal Historian of Oz [Part One]." *Magazine of Fantasy and Science Fiction* **8**, 1 (January 1955) 71-81.

   - "The Royal Historian of Oz [Part Two]." *Magazine of Fantasy and Science Fiction* **8**, 2 (February 1955) 64-74.
   - *O&S* (1983) 115-137.

     Combines the text of the two articles, with new postscript

   - *TNIL* (1996) 328-347.

     Combines the text of the two articles, with new introduction and postscript

2. "John F. Snow, 49, Continued Oz Books." *New York Times* (July 14, 1956) 15.

   Gardner wrote this obituary of Jack Snow, with no by-line.

3. "A Few Words from our Chairman." *Baum Bugle* **1**, 1 (1957) 4.

   - "A Few Words from our Chairman." *Best of Baum Bugle: 1957–1961* (International Wizard of Oz Club, February 1966) 6.
   - "A Few Words from our Chairman." *Baum Bugle* **41**, 1 (Spring 1997) 5.

4. "The Librarians in Oz." *Saturday Review* **42**, 15 (April 11, 1959) 18-19.

   - *O&S* (1983) 138-141.

5. "Why Librarians Dislike Oz." *American Book Collector* (December 1962) 14-16.

   - "Why Librarians Dislike Oz." *Library Journal* **88** (February 15, 1963) 834-836.
   - "Why Librarians Dislike Oz." In *The Wizard of Oz (The Critical Heritage Series)*, edited by Michael Patrick Hearn (Schocken Books, 1983) 187-191.
   - *O&S* (1983) 142-146.

6. "Oz Revisited." *Baum Bugle* (Spring 1964) 15.

   - "Oz Revisited." *Best of Baum Bugle: 1963-1964* (International Wizard of Oz Club, 1975) 50.

7. "The Land of Oz, U.S.A.." *Baum Bugle* (Spring 1965) 9-10.

   - "The Land of Oz, U.S.A.." *Best of Baum Bugle: 1965-1966* (International Wizard of Oz Club, 1982) 11-12.

8. "A Child's Garden of Bewilderment." *Saturday Review* (July 17, 1965) 18-19.

   - "A Child's Garden of Bewilderment." In *Only Connect*, edited by Sheila Egoff, et al (Oxford University Press, 1969) 150-155.
   - *O&S* (1983) 164-168.

9. "John Dough and the Cherub." In *Children's Literature: The Great Excluded (Vol. 2)*, edited by Francelia Butler (Children's Literature Assoc, 1973) 110-118.

   Same text as the introduction to the book *John Dough*.

   - *JFH* (2008) 221-233.

10. "[Thoughts on the MGM movie]." *Baum Bugle* (Autumn 1989) 7-8.

11. "*The Enchanted Island of Yew*: Baum's Adult Children's Book." *Baum Bugle* (Spring 1990) 13-14, 29-30.

    States it will be an introduction to a (unpublished) Dover edition.

    - *JFH* (2008) 213-220.

12. "*The Tin Woodman of Oz*: An Appreciation." *Baum Bugle* (Fall, 1996) 14-19.

    - "The Tin Woodman of Oz." In *UTB* (2003) 164-173.

13. "Mother Goose in Prose." *Baum Bugle* (Winter 1997) 8-12.

    - *JFH* (2008) 251-258.

14. "Word Play in the L. Frank Baum Fantasies." *Word Ways* (May 1998) 137-138.

- "Word Play in the L. Frank Baum Fantasies." *Baum Bugle* (Autumn 1998) 26-28. illustrated.
- *WJ* (2000) 97-103.
- "Word Play in the L. Frank Baum Fantasies." *Word Ways* **43**, 3 (August 2010) 224-225.

15. Martin Gardner and Ruth Berman, "Frederick Richardson." *Baum Bugle* (Spring 1999) 15-27.

The "introduction" to this is taken from the Dover *Queen Zixi of Ix*; the remainder is by Ruth Berman; illustrated.

16. "How the Oz Club Started: and What Happened Next." *Baum Bugle* **51**, 2 (Autumn 2007) 10-12.

- *JFH* (2008) 259-262.

15.4 Book Reviews

1. "Follow the Yellow Brick Road." *New York Herald Tribune Book Week* (January 2, 1966) 14.

The Wizard of Oz, L. Frank Baum; the other Baum Oz books are mentioned.

2. "We're Off to See the Wizard." *New York Times Book Review, Part II, Children's Books* (May 2, 1971) 1, 41.

The Annotated Wizard of Oz, M.P. Hearn

- *O&S* (1983) 272-276.

3. "The Master Key: An Electrical Fairy Tale." *Baum Bugle* (Autumn 1974) 22.

The Master Key, Lyman Frank Baum

4. "Santa Was an American." *New York Times Book Review* (December 7, 1975) 8, 12.

The Life and Adventures of Santa Claus, Lyman Frank Baum

- *O&S* (1983) 169-172.

15.5　Published Correspondence

1. "Oz & Ends." *Baum Bugle* (Christmas 1964) 17.

   Mentions merchandise that was available.

2. "Letters." *Scientific American* **233** (November 1975) 8.

   Points out that some bacteria have 'propellers' like the Ork of Oz.

3. "A Feminist's Legacy." *Free Inquiry* (Spring 1994) 63.

   Discusses the connection between feminist Matilda Gage and *The Land of Oz*.

4. "The Wizard of Oz and Madame Blavatsky." *Chesterton Review* **31**, 1/2 (Spring/Summer 2005) 253.

   Discusses the connection between theosophist Matilda Gage and *The Land of Oz*.

Juvenile Literature

In the Magic Chapter we see that in the early 1940s he wrote two booklets for youth about science, using fun science tricks. After the war he had a monthly fun column in *Uncle Ray's Magazine*; see the Math Chapter. In the early 1950s, he began contributing regularly to Parents Institute, which was keen to expand their magazine line. In a short time he was an employee, often working from home. This was one the rare times in his life when he was not a free-lancer. He edited or contributed to six titles for children. He was principally associated with *Humpty Dumpty*, which he edited from the first issue. He told his mother he was the "contributing editor in charge of gimmicks." He provided filler material, games, stunts, puzzles and the occasional article. For Humpty Dumpty he also contributed a short story about the titular character and he also contributed a cautionary verse about good behavior. Some of the poems were collected in *Never Make Fun of a Turtle, My Son*, but the short stories have never been republished despite his concerted efforts.

He never was evasive about his work at Parent's Institute. He was proud of it. He continued to write for the juvenile market long after that.

16.1 Books

1. *The Arrow Book of Brain Teasers* (Tab Books / Scholastic Book Services, 1959).

    From Parents Institute periodicals

2. *Science Puzzlers* (Viking Press, 1960).

    - *Science Puzzlers* (Scholastic Book Services, 1966).
    - *Entertaining Science Experiments with Everyday Objects* (Dover, 1981).
    - *Searching Science* (New Delhi: Learners Press, 1998).
    - *Exploring Science* (New Delhi: Learners Press, 1998).

    These two are a pirated edition issued as two volumes.

3. *Mathematical Puzzles* (Crowell, 1961).

    - *Entertaining Mathematical Puzzles* (Dover, 1986).

4. *Archimedes, Mathematician and Inventor* (MacMillan, 1965).
5. *Perplexing Puzzles and Tantalizing Teasers* (Simon and Schuster, 1969).

    - *Perplexing Puzzles and Tantalizing Teasers* (Archway, 1971).
    - *Perplexing Puzzles and Tantalizing Teasers* (Dover, 1988).

    Bound together with *More Perplexing Puzzles and Tantalizing Teasers*.

6. *Never Make Fun of a Turtle, My Son* (Simon and Schuster, 1969).
7. *Space Puzzles* (Simon and Schuster, 1971).

    - *Space Puzzles* (Archway, 1972).
    - *Puzzling Questions About the Solar System* (Dover, 1997).

    "Revised and enlarged edition"

8. *Codes, Ciphers and Secret Writing* (Simon and Schuster, 1972).

    - *Codes, Ciphers and Secret Writing* (Archway, 1974).
    - *Codes, Ciphers, and Secret Writing* (Dover, 1984).

9. *The Snark Puzzle Book* (Simon and Schuster, 1973).

    - *The Snark Puzzle Book* (Prometheus, 1990).

10. *More Perplexing Puzzles and Tantalizing Teasers* (Archway, 1977).

    - *Perplexing Puzzles and Tantalizing Teasers* (Dover, 1988).

    Two volumes bound as one.

11. *Classic Brainteasers* (Sterling, 1994).

    Based largely on the *Arrow Book of Brainteasers*.

- *Brainteasers* (Sterling, undated).

  Small book in the "Pocket Puzzlers" series; contains selected problems from *Classic Brainteasers*, as well as from three books by other authors (Robert Steinwachs, George J, Summers, and Edward J. Harshman).

- *The Funtime Riddles and Trivia Book* (Sterling, undated).

  Booklet issued by The Sorrento Cheese Company (before 2003); contains selected problems from *Classic Brainteasers*, as well as from a riddle book by Joseph Rosenblum.

12. *Science Magic* (Sterling, 1997).

    Drawn from the column in *Physics Teacher*, with new material; see chapter on Physics Tricks for more detail.

    - *Science Tricks* (Sterling, 1998).

      Title change for the paperback edition.

    - *Martin Gardner's Science Magic* (Dover, 2011).

13. *Mind-Boggling Word Puzzles* (Sterling, 2001).

    - "Mind-Boggling Word Puzzles." In *MENSA Word Puzzles* (Main Street / Sterling, 2004).

      Packaged with similar books by Mark Danna and Helene Hovanec. No reason for MENSA label.

    - "Mind-Boggling Word Puzzles." In *The World's Biggest Book of Brainteasers and Logic Puzzles*, edited by Bea Kimble (Sterling, 2006).

      Packaged with similar books by Willis, Sloane, MacHale, Dispezio, Smith, Sole and Marshall.

    - *Mind-Boggling Word Puzzles* (Sterling, 2007).

      Small 24 page abridgment, with "MindWare" on cover; presumably distributed with MindWare products.

    - *Mind-Boggling Word Puzzles* (Dover, 2010).

14. *Smart Science Tricks* (Sterling, 2004).

    Drawn from the column in *Physics Teacher*, with new material; see chapter on Physics Tricks for more detail. "Thanks too to Glen Vecchione, for his masterly assistance with the science explanations."

    - *Smart Science Tricks* (Goodwill Publishing House, 2008).

15. *Optical Illusion Play Pack* (Sterling, 2008).

16.2   Books Edited

1. [George Carlson], *Peter Puzzlemaker: A John Martin Puzzle Book for Little Puzzlers* (Dale Seymour, 1992).

   Illustrated by George Carlson, who presumably should be listed as the author. Gardner's fond memories of this book led him to get it reprinted. The book was intended to be used by teachers to make transparencies.

2. [George Carlson], *Peter Puzzlemaker Returns* (Dale Seymour, 1994).

16.3   Columns and Periodicals Edited

The citations below (often just filler material) are hit and miss, as most of them are uncredited. Therefore we rely on tear-sheets found in Gardner's files (which he stated were all by him), saved in case he wanted to reuse them. Instead Parents Institute decided to reprint them, without consulting Gardner, starting around 1960, often in completely different magazines. No attempt is made to track down reprint citations, since they were far-flung. Surprisingly *Children's Digest* in 1971 reprinted material from that magazine but credited it instead to Gardner's *Arrow Book* where they had been collected!

1. *Children's Digest* **1**, 10 (September 1951).

   Gardner was a contributing editor until 1961. He mainly contributed tricks, riddles, puzzles, and "brain teasers" as filler material. Occasional this material was expanded into a separate piece. If a piece below appeared with a by-line (on either the article and/or the Table of Contents) it has an asterisk (*). Many appeared without attribution but are from tear-sheets in his files, often undated and without page numbers.

   Articles
   "Baffling Baseballs," 1951
   "Blow Your Own Horn," October 1951, 44-45 *
   "Change a Lion into a Bear"
   "The Fortune-telling Witch," 127
   "Goofy Arithmetic"
   "Hidden Names," December 1953

Games

Fun with ...

Puzzle Parade and Puzzle Corner

"The Astronomers / Lost in the Woods / Sheer Mistake," July 1958, 44-45, 77

"Baby Checkerboard"

"Ben Franklin at Home"

"Broken Chain"

"Perplexed Scouts / Crazy Timepiece / Three Little Strips / Four Jolly Players / Guess the Letters / Find the Missing Word," January 1955, 72-74, 130

"Fair and Square," April 1955

"Five Fishy Questions," September 1955

"Geography Rebus / Cross the Z's / Jane's Patchwork Quilt," December 1955, 98-99, 130

"Grave Error"

"Guess the Letter / Find the Missing Word / Perplexed Scouts," 98-99, 130

"Low Bridge / High IQ"

"Mixed-up Mailman / Tricky Fish / Penny Puzzle / Alphabet Zoo," etc., February 1953, 44-45, 130

"Mysterious Tracks"

"Name the Men / Name the Animal / Name the Relation," September 1957

"Odd Trip / Lost Star / Soprano Solo," April 1957

"Pepper or Salt / Puzzling Grocer," April 1955

"Perplexing Scissors"

"Rearrange the Pins / Add That Line / State Those States," etc., 44-45, 77

"Slice the Pretzel"

"Square the Cross," April 1954

"Test Your 'I' Q"

"Three Paths / New Pitcher / Sports Rebus," April 1954

"Triangular Cat"

"Up and Down Exchange"

Science Corner

"Color Magic (disks) / Mysterious Bottle," February 1955

"Defy Gravity (Bernoulli) / Looking Through a Tube with One Eye," March 1954

"Make Your Own Planetarium," February 1955, 62-65

"Musical Spoon / Balanced Bottle / Rainbow," April 1953

"Paper Cup Telephone / Twisty Snake / Magic Print Shop [lifting cartoons]," October 1953

Science is Fun
"Balanced Yardstick / Floating Button," January 1957
"Curious Cork / Draw Ellipse," April 1959
"Disappearing Bunny / Talking Machine," October 1957
"Elusive Dollar Bill / Three Streams," January 1958
"Homemade Movies / Siphon String," April 1956
"Milk Carton Guitar / Floating Needle," October 1955
"Salt-Ice Cube / Floating Egg," May 1957
"Timothy's Magic Name," September 1960
"Unreversed Image," December 1959

Science Bloopers
"A Major Mistake," May 1959, 77, 117

Tricks to Try
"Dissolving Knots," May 1958
"Empty Cup," March 1960
"Great Card Mystery," October 1956
"The Great Word Mystery," February 1958, 88-91
"Magic Key," May 1956
"Magic Rabbit," May 1959, 102-105
"The Mystery of the Linking Clips," November 1957
"Ring on a String," July 1955
"Swinging Cups," December 1958
"Vanishing Cup," March 1957
"Vanishing Pencil," November 1955

Magic Corner
"Boomerang Hat / Enchanted Crayons," May 1954, 76-78
"Multiplying Pennies / Mysterious Box," Feb. 1954, 110-112
"Vanishing Dime / Obedient Ring," September 1953, 10-12

Filler Items
"Brain Teasers," February 1958, 36, 97
"Brain Teasers," July 1958, 24, 77
"Count the Squares"
"Crazy Crossword"
"Crossword Puzzle," 77, 130 *
"Donald Gets His Skates," November 1955, 26
"Draw Some Crumply Profiles," 130
"The E-Z A-Z Game," December 1956, 76

"Fifi Takes a Walk," October 1954, 38
"Finger Fun," April 1956
"Fun with Words"
"Games for Rainy Days," December 1952, 34
"The Goose or the Egg," November 1953, 127, 130
"The Great Hat Swindle," May 1953, 52
"Hidden Names," December 1953, 26
"Hidden Proverb," February 1955, 37
"How Tough is an Egg," September 1954, 16
"Knots or Not"
"Let's Play Umbrella Bounce," March 1956, 110
"The Letter that Opens Itself," 13
"This Looks Easy Until You Try It," October 1952, 65
"Make a Button-Spinner," December 1951, 99
"Magic Square" (not a magic square), October 1953, 11
"Match Your Map Wits," May 1954, 36
"Mysterious Licenses," October 1953, 121
"Pitch a Game of Bowls," September 1954, 104
"Play Calendar Toss," May 1956, 106
"Play Peanut Puff," December 1954, 130
"Play Touch and Guess," July 1955, 76
"Prove your Superstrength," November 1954, 80
"Raining Outside? Play Stairsteps," October 1951, 99 *
"A Revolutionary Puzzle," February 1954, 36
"Scented Sentence," May 1954
"The Square Pig that Looks Around," 41
"State the States," February 1956, 130
"Think of a Number," January 1956, 130
"The Three Daughters," April 1954
"Through the Dots," February 1955, 22
"Toothpick Tussle," July 1956, 56.
"Triangular Turkey," November 1957, 96
"The 21 Puzzle," February 1954, 72, 130
"Which Line is Longer," 85
"Wiggles," 95
"Wise Old Owl," January 1954, 12

2. *Parents' Magazine* (1952).

Gardner was a contributing editor for the "Family Fun department" which featured many entertaining home projects. For the first year and a half he received no or little credit; after that he had a by-line. Only the starting pages are given below.

1952 (Jan, 8; Feb, 83; Mar, 82; Apr, 102; May, 96; Jul, 78;
    Aug, 83; Oct, 122; Dec, 101)
1953 (Jun, 95; Jul, 80; Aug, 52; Sep, 56; Oct, 56; Nov, 56;
    Dec, 56)
1954 (Jan, 68; Feb, 52; Oct, 54)
1955 (Jan, 44; Feb, 106)

3. *Humpty Dumpty's Magazine* **1**, 1 (October 1952).

Listed is a story involving Humpty Dumpty Jr. (HDJ) and a
poem of cautionary verse (ended October 1957) (there were no
June or August issues); at times one or the other was missing.

1952
October – The Recent Adventures of HDJ 37-46 / – [no poem]
November – The Christmas Tree 50-57 / Do Not Open Until
Christmas 100-101
December – /
1953
January – The Talking Snowman 86-93/It's Snowball Time 48-49
February – The Unhappy Hippo 37-42 / Never Tell a Lie 16-17
March – /
April – /
May – The Bear That Overslept 100-107 / Never Make Fun 94-95
July – Mr. Goat Gives a Prize 86-93 / Table Manners 114-115
September – Junior Gets Hooked 70-77 / Play Quietly 78-79
October – /
November – /
December – /
1954
January – Saved by a Nose 72-79/ Don't Be a Show-Off 64-65
February – The Mixed-Up Valentines 105-110/ Are You a Gum-
Bug 62-63
March – Junior Loses a Tooth 112-119/ Sit Still! 34-35
(also "Feeny, the Funny Fireman," 57-66)
April – Junior's Easter Suit 102-109 / April Showers 70-71
May – The Caterpillar Who Tried to Fly 28-35/Be Polite 100-101
July – The Pony Who Wanted to Be a Cowboy 8-15 / Don't Be
a Scribble Scamp 90-91
(also "Feeny, the Funny Fireman," 100-109)
September – Junior's Good Deed 28-35/Sharing Your Toys 18-19
October – Herman Gets Stuck 80-7/Hang up You Clothes 118-9
November – The Big Parade 100-109 / Careless Feet 18-19
December – Junior's Christmas Sleigh Ride 8-16 / Don't be a
Chistmasnooper 76-77

1955
January – The Man from Mars 114-121 / Clean Hands 62-63
February – The Missing Walnuts 76-83 / Cry-Baby 92-93
(also "Feeny, the Funny Big Game Hunter," 108-118)
March – The Little Lost Mouse 24-31 / Spring Fever 102-103
April – Shoe Shine 74-81 /Sweet Tooth 102-103
July – Junior, the Great 8-15 / Take Your Medicine 108-109
September – Junior Learns to Swim 46-53 / Visit to the Dentist
120-121
October – Halloween Night 8-15 / Ring Around Junior 46-47
November – /
December – Bascom, the Blue-Nosed Bear 50-57 / –
1956
January – Davy Cricket 22-29/ Pretending 74-75
February – The Big Test 26-33 / Riding with Dad 56-57
March – Biff, Boff and Burp 80-87 / The Magic Word 20-21
April – Junior's Big Joke 90-97 / How Big? 16-17
May – Hocus Pocus 28-35 / Good Sport 22-23
July – /
September – Ping-Pong 104-111 / The Giggles 16-17
October – The Dancing Horse 38-45 / Copy Cat 108-109
November – The Thankful Moth 24-31/Don't Disturb Dad 48-49
(also "Feeny, the Funny Pilot," 84-93)
December – Junior's Secret 5-13 / Two of a Kind 108-109
1957
January – Hide and Seek 86-93 / January 96-97
February – Junior Bakes a Cake 78-85 / Speak Clearly 76-77
March – Chief Sitting Bull Stands Up 5-12 / March Wind 72-72
April – The Lion Tamer 98-105 / Think Twice 34-35
May – Junior, the Storekeeper 104-111 / –
July – The Bare-Headed Whale 24-31 / –
September – Junior Builds a TV Set 44-51 / –
October – Birthday Dinner 5-13 / Party Smarty 68-69
November – The Pig Who Went to Market 5-13 / –
December – /
1958
January – Dr. Junior 88-95 / –
February – / –
March – Organ Grinder 24-31 / –
April – The Stolen Emerald 100-107 / –
May – Joe Mink 66-73 / –

July – The Showdown 46-53 / –
(also "Feeny, the Funny Fisherman," 90-99)
September – / –
October – The Lady and the Dragon 102-109 / –
November – Laugh Bird Laugh 48-55 / –
December – / –
1959
January – / –
February – / –
March – Easter Hat 6-13/ –
April – / –
May – / –
July – Summer Picnic 6-13/ –
September – The Red Socks 44-51 / –
October – /
November – A Turkey for Thanksgiving 42-49 / –
December – Punching Bag 6-13 / –
1960
[nothing in 1960]
1961
January – Alexander Graham Baboon 66-74 /
February – The Man from Mars 90-97 (first reprint, from January
1955) / –
March – Smoky Mountain 68-75 / –
April – The Dancing Robot 40-47/ –
May – The Caterpillar Who Tried to Fly 28-35 (May 1954)/ –
July – Billy the Kid 102-109 / –
September – Beach Ball 88-95 / –
October – / –
November – The Big Parade 28-37 (November 1954)/ –
December – / –
1962
January – HDJ stories continued by Jay Williams/

4. *Polly Pigtails* **1**, 1 (Spring 1953).

   Gardner was the managing editor until February 1954. Contributed "Polly's Party Pranks", Polly's letter, and various fillers.
   The magazine was later called *Calling All Girls*.

5. *Piggly Wiggly* **1**, 1 (Winter 1953).

   Gardner was the managing editor for the first two issues; the
   second was Spring 1953. Beginning with the third issue the magazine changed its name to *Piggity's*. He contributed a story about

Piggly Wiggly / Piggity to each of the first four issues. He also contributes activities and puzzles. In March 1954 it was combined with *Humpty Dumpty*.

"How Piggly Wiggly Became a Champion," Winter 1953, 5-13
"Piggly Catches a Whale," Spring 1953, 5-14
"How Piggity Became a Fireman," Summer 1953, 5-13
"[Unknown]," Fall 1953

6. *Children's Playcraft* (January 1954).

Gardner contributed activity features and fillers until Summer 1954. These appeared without attribution but are from tearsheets, often undated, in his files:

"Bet You Can't," January 1954, 68
"Bet You Can't," March 1954, 68
"Bet You Can't," Summer 1954, 68
"Electrify Your Friends,", No.2, February 1954, 55-59
"The Floating Paper," 100
"4 Balloontrix," January 1954, 92-93
"Fun with Shoeboxes"
"Fun with Snow"
"Give a Space Party," February 1954, 42-46
"The Great Box Puzzle" (Towers of Hanoi), March 1954, 112-114
"The Great Winkletum," Summer 1954, 20-23
"A 'Handy' Calendar," 69
"Jocko the Climbing Monkey"Summer 1954, 114-116
"Juggling Made Easy," February 1954, 5-7
"Jumpo," January 1954, 102-103
"Let's Go Bathtub Fishing,",69
"Let's Play Indoor Horseshoes," 42-43
"Let's Play Wordo," February 1954, 16-17
"Let's Play Zoograms" (tangrams), January 1954, 54-55, 84
"Make a Baby Boomerang"
"Make a Jumping Jack Clown"
"Make a Magic Bug Catcher," Summer 1954, 85-89
"Make a Mystery Spinner"
"Make Your Own," codes

"The Mysterious Telephone Code,"
"Penny and Dime Puzzle," January 1954, 12-13
"Play Rocket to the Moon," Summer 1954, 100-101
"Scramble and Unscramble,"
"The Secret of the Baffling Belts" (Moebius strips), February 1954, 92-95
"Send Secret Messages,"January 1954, 82-83
"Seoda Straw Fun," Summer 1954, pp, 5-10
"The Secret Triangle Code,"
"6 Tricks with Inertia," Summer 1954, 39-42
"Toothpick Teasers,"February 1954, 40-41
"The Yogi Tells All," January 1954 32-35

16.4   Articles

1. "Edison: Wizard of Light." *Children's Digest* (November 1954) 89-96.
2. "Your Senses Can Fool You." *Children's Digest* (December 1955) 33-39.
3. "The Whirling Worlds of Outer Space." *Children's Digest* (November 1956) 37-47. [No by-line]
4. "IGY: Science's Greatest Challenge." *Children's Digest* (November 1957) 33-40.
5. "Fun with Gravity." *Children's Digest* (July 1958) 25-31.
6. "Fun with Sound." *Children's Digest* (October 1958) 33+.

16.5   Anthologized Material

1. *Humpty Dumpty's Album for Little Children* (RCA Victor (LBY-1015), 1958).

   Vinyl record of readings of several selections from the magazine:
   "January "
   "Copy-cat Twiddly Bug"
   "Party Smarty"
   "Think Twice"
   "Humpty Dumpty Jr. Builds a TV Set"

2. "Never Make Fun." In *Our English Language*, edited by M. Bailey et al. (American Book Co., 1956) 197.

3. "Shoe Shine." In *Humpty Dumpty's Storybook* (Parents' Magazine Press, 1966) 66-72.
4. "Bascom the Blue-Nosed Bear." In *Humpty Dumpty's Holiday Stories* (Parents' Magazine Press, 1973) 9-16.

   Also contains "Junior's Big Joke", pp. 32-38.

5. "Never Make Fun." In *Never Give Up*, edited by E.L. Evertts (Holt, Rinehart and Winston, 1973) 35.
6. "[Various Poems]." In *Crickets Choice*, edited by Clifton Fadiman and Marianne Caras (Open Court, 1974).

   The magazine *Cricket* reprinted many puzzles from the book *Science Puzzlers* some of which are collected here:
   Dime Illusion, p. 131
   Floating Frankfurter, p. 188
   Missing Slice, p. 278

7. "[Various Poems]." In *Humankind*, edited by Helen Weber and Marianne Caras (Open Court, 1976).

   Another collection of science puzzlers from *Cricket* magazine published as a series of modules:
   Level C ("The Egg and Inertia", p. 35; "Through the Hole", p. 54)
   Level D ("Floating Frankfurter", p. 14; "Missing Slice", p. 15)
   Level E ("Two Science Experiments", pp. 28-29)

8. "Cry-Baby." In *I'm Mad at You*, edited by William Cole (Collins, 1978) 15.

   Also contains, "Magic Word", p. 28.

   - "Cry-Baby." In *I'm Mad at You*, edited by William Cole (Random House, 1979) 15.

9. "Speak Clearly." In *Morning, Noon, and Nighttime, Too*, edited by L.B. Hopkins (Harper and Row, 1980) 38.
10. "Magic Word." *School Magazine* [New South Wales Department of Education] **65**, 7, Part 2 (August 1980) 218.

    Also contains, "Cry-Baby", p. 196, in Part 4.

11. "Speak Clearly." In *Poem Stew*, edited by William Cole (Harper and Row, 1981) 32.
12. "Cry-Baby." In *The Beaver Book of Funny Rhymes*, edited by Barbara Ireson (Beaver, 1990) 9.

13. "Speak Clearly." In *Poem Stew*, edited by William Cole (Lippincott, 1981) 32.
14. "Barbershop." In *The Random House Book of Poetry for Children*, edited by Jack Prelutsky (Random House, 1983) 113.

    Also contains, "Soap", p. 138.

15. "Soap." In *Ourselves and Others* (Graphic Learning Corporation, 1984) 42.

    Also contains, "Barbershop", p. 68.

16. "The Giggles." *Instructor* **94**, 2 (September 1984) 67.
17. "Magic Word." In *Open Roads* (Beka Books, 1985) 189.
18. "Soap." In *Reading Literature Workbook: Gold Level* (McDougal, Littell and Company, 1989) 118.
19. "Barbershop." In *Spelling and Vocabulary: Level 3 (Teacher's Edition)* (Houghton Mifflin, 1990) 124.
20. "The Giggles." In *For Laughing Out Loud*, edited by Jack Prelutsky (Knopf, 1991) 72.
21. "Magic Word." *Clubhouse Junior* **5**, 5 (1991) 2.
22. "Barbershop." In *Favorite Poems of Childhood*, edited by Philip Smith (Dover, 1992) 63.
23. "Soap." *Ladybug* (May 1995) 35.
24. "Soap." In *The Fish Is Me: Bathtime Rhymes*, edited by Neil Philip (Houghton Mifflin, 2002) [3].
25. "Soap." In *Family Poems*, edited by Belinda Hollyer (Kingfisher [Houghton Mifflin], 2003) 64.
26. "Speak Clearly." In *The Bill Martin Jr. Big Book of Poetry*, edited by Bill Martin Jr and Michael Sampson (Simon and Schuster, 2008) 134.
27. "[Various Poems]." *Word Ways*.

Poems from *Humpty Dumpty* were often used as filler material:

"Pretending," August 2008, vol. 41, no. 3, p. 203
"Never Make Fun," August 2008, vol. 41, no. 3, p. 205
"Share Toys," August 2008, vol. 41, no. 3, p. 238
"The Giggles," February 2009, vol. 42, no. 1, p. 13
"Medicine," February 2009, vol. 42, no. 1, p. 45
"Cry Baby," February 2009, vol. 42, no. 1, p. 51
"Riding with Mom," May 2009, vol. 42, no. 2, p. 95
"Scribble Scamp," May 2009, vol. 42, no. 2, p. 132
"Magic Word," August 2009, vol. 42, no. 3, p. 197
"Politeness," November 2009, vol. 42, no. 4, p. 254

"Good Sport," November 2009, vol. 42, no. 4, p. 258

"Autumn," November 2009, vol. 42, no. 4, p. 289

"Slowpoke," August 2010, vol. 43, no. 3, p. 206

"Barbershop," February 2011, vol. 44, no. 1, p. 2

"Bedtime," February 2011, vol. 44, no. 1, p. 13

"Twiddlebug," February 2011, vol. 44, no. 1, p. 81

"Supermarket," May 2011, vol. 44, no. 2, p. 86

"Party Smarty," May 2012, vol. 45, no. 2, p. 84

"Careless Feet," August 2012, vol. 45, no. 3, p. 164

"January," November 2012, vol. 45, no. 4, p. 244

Figure 16.1: Brittle clipping as found in Gardner's files.

Chapter 17

Journalism

In all his writing Martin Gardner described himself as a journalist, simply because he often was explaining what someone else had done. However there were times when he actually worked in the newsroom or edited copy. His journalistic experience at the University of Chicago, led him to be a yeoman in the Navy and to be asked to edit their newspaper at the radio training school. This environment of deadlines helped him in later life. He never suffered writer's block.

17.1   Periodicals Edited

1. *Comment* [University of Chicago] **4**, 1-4 (November 1935).

   Edited one volume of this "Literary and Critical Magazine".

2. *The Badger Navy News* 1 (January 16, 1943).

   Edited and wrote all material that appeared without bylines. It was an insert in the *The Daily Cardinal* (University of Wisconsin - Madison) that was "Published for the University's Naval Training School." Often 2 pages long. It continued until number 72, November 2, 1944.

3. *The Wisconsin Alumnus* (1943-1944).

   As an extension of his responsibility to produce *The Badger Navy News* he also contributed Navy news to this publication and at one point (October 1943) he appeared on the masthead as the "Navy Editor".

"Naval Training School," vol 44, no 4, July 1943, pp. 326-327.
"The Navy," vol 45, no 1, October 15, 1943, p. 10.
"The Navy," vol 45, no 2, November 15, 1943, p. 40.
"The Navy," vol 45, no 3, December 15, 1943, p. 73.
"The Navy," vol 45, no 4, January 15, 1944, p. 106.
"The Navy," vol 45, no 5, February 15, 1944, p. 10.
"The Navy," vol 45, no 6, March 15, 1944, p. 10.
"The Navy," vol 45, no 7, April 15, 1944, pp. 8-9.

4. *Inside New York* (March-April, 1950).

Edited and wrote some material that appeared without bylines.
It was "A newsletter issued by the New York office [of the] Amer-
icans for Democratic Action." It was a small 4-page newspaper.
Unknown duration; only two issues known.

17.2 Articles

Some citations are from cuttings without dates and/or page numbers.

1. "Yes, Life Is Like That in Goldfish Bowl." *Tulsa Tribune* (January 10, 1938).
2. "Doctoring Well with Large Capsules Solves Water Source." *Tulsa
   Tribune* (January 23, 1938). no byline
3. "Orientation of Cores an Accomplished Feat." *Tulsa Tribune*
   (February 13, 1938).
4. "Versatile Hypnotist Here; He's Done Nearly Everything." *Tulsa
   Tribune* ([1938]).
5. "Seismograph Was Constructed More Than 1800 Years Ago."
   Tulsa Tribune ([1938]).
6. "Geophysics Aid to Exploration." *Tulsa Tribune* ([1938]).
7. "Device in Tulsa Checks Records Over the World." *Tulsa Tribune*
   (April 17, 1938).
8. "Famous Magician Stumped by Oil Exposition 'Magic'." *Tulsa
   Tribune* (May 18, 1938). no byline
9. [Martin Gardner], "Fraudulent Gambling Devices." *The Chicago
   Reporter* **1**, 2 (January 18, 1939) 2-3.
10. [Martin Gardner], "Crime: Evasion?" *The Chicago Reporter* **1**,
    3 (February 1, 1939) 1,4.
11. "News of the Quadrangles." *University of Chicago Magazine*
    (January 1942) 14-17.

Miscellaneous

"Miscellaneous" is self-descriptive. Unusual entries include his published caricatures, which were popular at the University of Chicago. Also included are his letters to local newspapers (collected here regardless of the topic). He considered such letters to be a civic duty.

18.1 Books

1. Anthony Ravielli [Martin Gardner], *An Adventure in Geometry* (Viking Press, 1957).

   Gardner ghosted this book for Anthony Ravielli, the illustrator.

2. Don Carter [Martin Gardner], *10 Secrets of Bowling* (Viking Press, 1958).

   Gardner ghosted this book for the intended ghost-writer, Anthony Ravielli, the illustrator. A "Revised Edition" was issued in 1963.

   - Don Carter [Martin Gardner], *10 Secrets of Bowling* (Cornerstone Library, 1961).

     "New Revised Edition" appeared before the "Revised Edition" above.

223

3. *Undiluted Hocus-Pocus: The Autobiography of Martin Gardner* (Princeton University Press, 2013).

Book was intended as a desultory memoir but was published posthumously as an "autobiography."

- Gardner, Jim, Kearns, Vickie, "On the Autobiography of Martin Gardner." In *G4G11 Exchange Book: Volume 2* (Gathering 4 Gardner, 2017) 8-30.

  Contains many pages of reproductions of the raw manuscript.

18.2   Pamphlets

1. *Lookin in on the Pope* ([privately published], [1946]).

Gardner YM2/c wrote the "History of the USS Pope" for this souvenir book for crew members.

18.3   Columns

1. Hidebrand and Henry, "This Side (and the Other)." *The Daily Maroon* [University of Chicago] (1934) 1.

An anonymous irregular column about the University of Chicago; each paragraph alternately written by Gardner and his friend, whose name was forgotten. Appeared November 6, 15, and 22.

2. "'Round the Oil Offices." *Tulsa Tribune* (December 26, 1937 – May 22, 1938).

Appeared every Sunday in the oil news section; chat garnered from visiting the oil company offices each week.

18.4   Book Introductions

1. Antonio Peticov, *Pinturas de 1993* (Galeria Nara Roesler, November 10 – December 4, 1993) 5.

This is the program for an exhibition of the artist Peticov in São Paulo. The introduction has only appeared in Portuguese.

2. John Wilcock, *Manhattan Memories: An Autobiography* (self-published, 2009) v.

18.5   Articles

1. "Believe It or Not." [Syndicated] (February 19, 1930).

   As many others had done, Martin Gardner sent a unusual message to Ripley to see if it would be delivered. There is a drawing of a postcard with just a postmark, "Tulsa Jan 4 1930," and the text "A postcard with no address but with a personal message on the back was delivered promptly to me." Gardner's name does not appear, but he saved the clipping. (Later the same year the Post Office stopped delivering mail that was not properly addressed.)

2. "Now It Is Now It Isn't." *Science and Invention* (April 1930) 1119.

   This was a question (# 2349) to "The Oracle" about the process of reversing the effect of ink eradicators.

3. "From Martin Gardner." *The Totem* (February 1932) 1-2.

   This is the Camp Mishawaka newsletter. He attended the Camp in the summers of 1927 and 1928, and reminisces about it. He includes a poem about the Camp and a few jokes.

4. "Coudich Castle: Bohemian Burrow From Way Back." *Phoenix* [University of Chicago] (March 1937) 6, 23.

5. "The New Economic Education." *Forecast for Home Economics* **57**, 8 (October 1941) 23, 64, 66, 68, 70.

6. "Alumni Book Exhibit." *The University of Chicago Magazine* (November 1941) 50-51. (No by-line)

7. "General Mills' Magic Campaign Revives Odd Field." *Advertising Age* (November 111, 1946).

   Uncredited article about Traub and Fun Inc.; Gardner's role mentioned.

8. "The Zaniest Firm in the Southwest." *Magazine Tulsa* **3**, 3 (March 1948) 32-33.

   About Roger Montandon

9. "Tulsa's Fabulous Bookman." *Magazine Tulsa* **3**, 4 (April 1948) 21-22.

   About John Bennett Shaw

10. "[Time Magazine Advertisements]." *New York Times* (January 26, 1961; February 23, 1961; March 8, 1961) 45; 28; 68.

   Three ads with games themes: bridge; checkers; tictactoe.

11. "Books They Liked Best." *Chicago Tribune* (December 3, 1967) S6.

   The solicited answer to the titular question was *A Personal Anthology*, by Jorge Luis Borges; *Nabokov: His Life in Art*, by Andrew Field; *Grooks*, by Piet Hein

12. In *The Meaning of Life*, edited by Hugh Moorhead (Chicago Review Press, 1988) 70.

   The solicited answer to the titular question was simply "I trust God knows, but not I."

13. "[Various]." In *The Reader's Catalog*, edited by Geoffrey O'Brien, et al. (Random House, 1989) 1068-1074.

   Gardner is a "consultant" to this listing of books. "I edited the many pages on mathematics." It includes a sidebar of his books.

14. "'My Grandmother is Blind': Memories of John Bennett Shaw." *Baker Street Journal* **40** (December 1990) 207-209.

15. "A Special Afterword from Martin Gardner." *Skeptical Inquirer* **32**, 5 (September/October 2008) 15.

   A note of thanks about the Eighth Gathering for Gardner

16. "Remembering John Bennett Shaw." *Friends of the Sherlock Holmes Collections* **13** (December 2009) 6. (Published by the Special Collections Library at the University of Minnesota.)

   - "Remembering John Bennett Shaw." *The Best of Friends*, edited by Rat Riethmeier, et al. (The Norwegian Explorers bof Minnesota, 2019) 93-94.

18.6   Book Reviews

1. "Off the Press." *The Daily Maroon* [University of Chicago] (October 2, 1936) 3.

   Winds over the Campus, James Weber Linn

2. *International Journal of Ethics* **48**, 4 (July 1939) 508-509.

   Ghandi Triumphant, Haridas Muzumdar
   Mahatma Ghandi, P.A. Wadia
   Lectures on Japan, Inazo Nitobe

3. "For Your Own Information." *Book Week – Paperback Issue* (January 710, 1965) 16-18.

   Nine desktop reference works.

4. "Petroski's Primer on the Pencil." *News and Observer* [Raleigh, NC] (January 7, 1990).

   The Pencil, Henry Petroski

   - *WW&FL* (1996) 206-208.

5. "From Papyrus to Pixels." *Civilization* (October/November 1999) 26.

   The Book on the Bookshelf, Henry Petroski

   - *WJ* (2000) 176-178.

18.7 Caricatures

1. *Phoenix* [University of Chicago] (1935).

   This humor magazine used Gardner caricatures and cartoons throughout the year, usually signed "Gard'r" or just "M.G.". Also see April 1936 and January 1937.

2. "[Various professors]." *Phoenix* [University of Chicago] (1937).

   Throughout the year this magazine used Gardner's large caricatures of various University of Chicago professors and students: Ronald Salmon Crane, Charles Morris, and Dennis McEvoy.

   - *Hypotheses: Neo-Aristotelian Analysis* 1 (Spring 1992) 1.
     This cover illustration reprints the Crane caricature. In Number 2, p. 24, a previously unused 1937 caricature of Mortimer Adler appears.
   - "Caricature of Charles Morris." *Semiotica* **23**, 1/2 (1978) 1-4.

3. *Pulse* [University of Chicago] (November 1940) 11.

   This "student magazine" used a caricature for an article on George Bernard de Huszar.

4. *The Foundation News* [University of Chicago Alumni Foundation] **1**, 5 (December 5, 1940) 1.
5. *The Ballad of Terrible Mike* ([2001]) privately printed.

   This 8-page pamphlet reprints a poem by Sam Hair that had appeared in the *Phoenix* in the 1930's, using two of the original illustrations.

18.8   Published Correspondence

1. *The Daily Maroon* [University of Chicago] (March 6, 1934) 2.

   Attacks the dogmatic relativism of the Social Science department.

2. "Letter of the Week." *Saturday Evening Post* (March 22, 1941).

   Passed on a 1905 letter about the *Post* found at the University of Chicago

3. Margery Goodkind, "Times Talkies." [Chicago]*Daily Times* (December 15, 1941) 27.

   Not correspondence but a man-in-the-street [Hyde Park] response (with photo) to the assertion "If a man rules and a wife obeys, 61 per cent of marriages would succeed."

4. "'Sure, Rockwell Has His Weaknesses ...'." *New York Times* (Apr 23, 1972) D19.

   Says attacks on Norman Rockwell are superficial. In the "Art Mailbag."

5. *Grump* 6 (April/May 1966) 12.

   Contributed a joke to this humor paper edited by Roger Price: "Speak softly and carry a big stick." [signed] Casanova (occupational therapist).

6. "The Publicity Machine Runs out of Gas." *Village Voice* **12**, 18 (January 16, 1967).

   It seems likely that this was written by another "Martin Gardner." It is about a party for the release of an Elia Kazan novel.

7. "Jovian Art." *Science News* **116**, 7 (August 18, 1979) 115.

   Tongue-in-cheek comment about Jackson Pollock and the surface of Europa.

8. *Logophile* **3**, 4 (1980) 26-27.

   About the artificial language Alwatu.

9. *Omni* (June 1984) 12.

   Tongue-in-cheek comment on a letter from Stanley Milgram.

10. *The Epistle* (March 1993) 3-4.

    Chatty letter about his recent books in "The Newsletter of the First Baptist Church of Montclair."

11. *Hypotheses: Neo-Aristotelian Analysis* 18 (Summer 1996) 40.

    Remarks about friends at the University of Chicago.

12. "Y2K." *Time-News* [Hendersonville, NC] (January 1, 2000) 6A.

    One paragraph, among many, about what the next century holds; Gardner mentions biological advances.

18.9   Letters to Local Newspapers

Gardner lived in retirement for more than 20 years in Hendersonville, North Carolina. During that time he frequently wrote letters to the editors of the local newspaper the *Time-News*. Sometimes it was about purely local issues, but more often it was about larger issues that affected the residents, such as medical fraud, the false-memory syndrome, and religious effrontery.

These citations are taken from Gardner's file of clippings, and page numbers are rarely available. Also, if the editor's title for the letter is not descriptive, a better description is given.

Oct 9, 1982   "Dislikes astrology" –
               argues against a horoscope column.
Oct 13, 1984   "No access this time" –
               remark on the Reagan/Mondale debate

Aug 31, 1985	"Locke, Rousseau not atheists" – corrects Mead Parce
Feb 22, 1986	"Thanks for help in the snow"
Apr 25, 1987	"A dream after reading" – questions Oral Roberts wealth
Oct 18, 1987	"Editorial showed balanced judgement" – approves of PTL coverage
Feb 6, 1988	"Physician's letter right on target" – about cataracts and diet
Mar 12, 1988	"Swaggart's ego needs curbing"
May 15, 1988	"Suggests reader poll on daily horoscope"
Oct 5, 1988	"What's Bush talking about, anyway?" – "a thousand points of light"?
Dec 26, 1988	"Don't send more money to televangelists"
Mar 25, 1989	"Brody column a disservice to readers" – about homeopathy
Jun 28, 1989	"Who wants Swaggart off the air?" – why hasn't Swaggart sued?
Sep 18, 1989	"Don't be fooled by diet scams" – attacks Cal-Ban ads
Dec 2, 1989	"Upset over Cal-Ban ads" – complains that the ads run as articles
Apr 3, 1990	"Offended by signs" – fundamentalist billboard, "Welcome to Toyota country"
May 4, 1990	"Angry at tobacco export efforts" – critical of Helms' efforts
Jun 16, 1990	"Flora Lewis almost had it right" – socialist/capitalist mixed economies
Aug 13, 1990	"Picture of Roseanne Barr on Flair front" – problem with caption
Aug 19, 1990	"Upset by Cal-Ban ads" – a further protest over these ads
Sep 30, 1990	"Letter rationalized trickle-down theory" – attacks supply-sider theory
Jan 27, 1991	"Offended by Thomas column" – Cal Thomas's attacks on secular science
Apr 18, 1991	"Some suggestions for the newspaper" – opinions on Doonesbury and columns
May 30, 1991	"Too much attention focused on sex" – the true JFK story is religion
Oct 25, 1991	"Cable company should drop Swaggart show"

Dec ??, 1991   "Questions how bomb was used" –
               why was the A-bomb first used on civilians
Jan 15, 1992   "Protests 'medical scams' ads" –
               ads for glasses to correct cataracts
Feb 16, 1992   "Would like to see horoscope gone"

Apr 5, 1992   "Critical of ads for 'worthless pill' " –
               article/ad for Food Source One
Jun 2, 1992   "Reader responds to chiropractor's letter" –
               warns about marketing
Jul 7, 1992   "Accusation was false Gardner says of letter" –
               'devout Mormon' derogatory?
Aug 26, 1992   "Tobacco, Quayle and family values" –
               Quayle's statement about exports
Oct 22, 1992   "A little humor making the rounds" –
               joke about Quayle, Bush and Clinton
Jan 17, 1993   "Editorial page marred by contrary opinion" –
               Cal Thomas ruins op-ed page
Feb 14, 1993   "Experience with blacks shows military will adapt" –
               gay GIs
Jun 15, 1993   "Untrained therapists plant false memories" –
               plug for FMS Foundation
Jul 11, 1993   "Satanic ritual abuse is mostly a myth" –
               reply to Dr. Sanborn
Sep 24, 1993   "Praises local restaurant for banning all smoking"
Dec 26, 1993   "Praises use of wire story on false memory syndrome" –
               plug for FMS Foundation
Jan 15, 1994   "Old testament verse condones smoking" –
               'she lighted off a camel'
Feb 5, 1994   "Earlier letter written tongue in cheek" –
               see *Word Ways* (May 1994, p. 88)
Feb 15, 1994   "Children had classic signs of false memories" –
               plug for FMS Foundation
May 30, 1994   "Paper should reject ads on bogus diet pills" –
               ads for Poloxonol
Jun 15, 1994   "Attack by Falwell on Clinton scurrilous" –
               videotape accuses Clinton of murder
Oct 7, 1994   "False memories spur huge jump in cases" –
               recommends books about FMS
Nov 6, 1994   "Homeopathic drugs have no real medicinal value" –
               response to local drugstores

Nov 12, 1994   "Attempts to confirm homeopathy all failed" –
               response to letter
Dec 7, 1994   "Homeopathic drugs have no effect at all" –
               response to two letters
Jan 12, 1995   " 'Remembered' abuse psychiatry's scandal" –
               recommends more books on FMS

Jun 20, 1995   "Ad's 'miracle drug' really just a placebo" –
               ads for ArthurItis
Sep 23, 1995   "Toms story recalls old bird proverb..." –
               Toms and the Bakkers
Dec 12, 1995   "Magazine article addresses amalgams" –
               gives reliable dental info
Dec 18, 1995   "In scientific circles, no flap over fluoride" –
               recommends books
Jan 3, 1996   "Lack of controversy, not lack opposition" –
               more on fluoridation
Jan 21, 1996   "Paper did readers service by running guest column" –
               more on fluoridation
Feb 8, 1996   "Loud music doesn't fit with good dining"
Feb 29, 1996   "Paper should refuse ads for bogus books" –
               book on vinegar therapy
Apr 3, 1996   "What brand of creation does writer believe" –
               evolution is a 'fact'
Apr 16, 1996   "Surprised paper ran story about store" –
               crystal therapy store
May 5, 1996   "Support of tobacco is Helms' hypocrisy" –
               thou shalt not kill'
May 12, 1996   "Most Christians accept evolution as God's method" –
               attacks misquotes
Jul 13, 1996   "Dole may be damaged by tobacco remarks" –
               no more addictive than milk?
Aug 5, 1996   "Paper should be force for enlightenment" –
               letters from fundamentalists
Sep 25, 1996   "Letter's purpose was decry intolerance" –
               fundamentalists are intolerant
Oct 15, 1996   "Let's hope Bakker not our only claim to fame" –
               recalls Roberts/Tulsa example
Jan 16, 1997   "Numerology book a blatant rip-off" –
               criticizes ad in *Parade*
Feb 4, 1997   "Black and Hoffman pussycats beside Jesus" –
               Jesus did attack hypocrites

Apr 28, 1997	"Paper should have rejected that ad" – criticizes large fundamentalist ad
May 11, 1997	"Word 'eunuch' has nothing to do with marital status" – self-castration in *Bible*
Aug 28, 1997	" 'Contact' is a film hopelessly marred"
Oct 20, 1997	"Responsible companies don't try to disguise ads" – ad for ARTH-RX
Jan 24, 1998	"Ads did not follow newspaper's policy" – yet another unlabeled ad/article
Apr 19, 1998	"Wants to know which theory writer supports" – what type of creationism?
May ??, 1998	"Would paper accept my ditty as well?" – was Christian poetry an ad?
Jun 28, 1998	"Are tornadoes, meteors God's way of punishment?" – Pat Robertson quote
Nov 26, 1998	"Great papers screen their correspondence" – fundamentalist letters
Dec 11, 1998	"Paper should stop running these letters" – fundamentalist letters
May 16, 1999	"Government should curb movie violence" – laws are the only effective means
Jun 8, 1999	"New Age infatuated with bogus science" – plugs *Skeptical Inquirer*
Jul 26, 1999	"Sound of silence is better at dinner"
Nov 25, 1999	"Ritual shows higher education's mind-set" – Aggie bonfires and priorities
Jan 7, 2000	"Hardcover book's price is a rip-off" – Almanac by mail was no bargain
Feb 26, 2000	"Jefferson's Bible leaves out miracles"
May 11, 2000	"Vinegar diet book has no health value" – response to an ad
May 24, 2000	"Old diet pill recycled under a new name" – Cal-Ban now is Fat-Stopper, etc.

The first two letters of 1994 were reprinted as "'Cause the Bible Told Me So", *Word Ways*, vol 27, May 1994, p. 88.

After moving to Norman, Oklahoma, in 2003 Gardner wrote to the *Norman Transcript*. (The first three are reprinted in *JFH* (2008) pp. 133-136.) (The fifth and sixth are reprinted in *T&F* (2009) pp. 223-225.)

Oct 1, 2005	"What about light, Dr. Sharp" –
	suggests further questions for a creationist
Jul 19, 2006	"Ad both hilarious, deplorable" –
	remarks about the Head On commercials
Oct 29, 2006	"Rush wrong about drugs" –
	critique of Limbaugh's homeopathy endorsement
Jan 13, 2008	"Wrong question" –
	does Romney believe the Book of Mormon?
Oct 28, 2008	"Is socialism a dirty word?" –
	distinguishes democratic socialism from communism
Nov 17, 2008	"Only in America" –
	a rejoinder about Socialism and Hayek

Part IV

Magic

In an interview Martin Gardner said "My main interest in magic is because it arouses a sense of wonder about the natural world; the universe is like a huge magic trick and scientists are trying to figure out how it does what it does."

This part represents Gardner's 81 years in the magic literature. Gardner's first and last publications were in magic journals, from 1930 to 2010. He was never (!) a performer. (He was a demonstrator for Mysto-Magic sets at Marshall Fields, in 1938 and 1939.) He was a constant and respected source of ideas for close-up magic, usually impromptu tricks. While Gardner's many tricks and effects are interesting to other magicians, they also provide non-magicians great insight into his creative side (which is often overshadowed by his reputation as a popularizer of other's work). And for this reason, the individual tricks will be listed within his books, so the full creative scope can be seen and the tricks can be cross-referenced to other locations.

I am not a magician and hence have limited access to the vast and arcane archives necessary to track down the many citations in the magic literature. Luckily I have been helped by Roger Montandon, Richard Kaufman, and Martin Gardner, among others. Even so, these difficulties prevent any claims of completeness. Much of the material in this part originally appeared as an appendix to *Martin Gardner Presents* (1993).

MAGIC

FOR THE
SCIENCE
CLASS

Prepared under the direction of MARTIN GARDNER, MAGICIAN

Chapter **19**

Magic for the Public

Most of Martin Gardner's efforts were for the closed magic community. Inevitably he found himself writing for the general public. There is a long-standing tension between revealing techniques and hiding techniques. Magicians thrive when the public does not share their secrets. If an effect is identified with its creator it is considered good form to not reveal anything. But with the passage of time and the inevitable variations and improvements the claim is weakened. As a result there are scores of books written by magicians for the public. Gardner only dabbled in that book market, but often wrote articles. Much of this falls into the category of parlor magic and stunts. In fact he made a series of TV commercials wherein he explained how to do simple tricks. The Dover book reprints material that originally appeared only in magic shops.

19.1 Books

1. *Martin Gardner's Table Magic* (Dover, 1998).

   Contains material from the five pamphlets (*Match-ic, 12 Tricks with a Borrowed Deck, After the Dessert, Cut the Cards, and Over the Coffee Cups*) that has been reorganized; also, some new material.

19.2 Pamphlets

1. *Magic for the Elementary Science Class* (Scott, Foresman, and Company, [1941]).

   Distributed to teachers using the textbook series *Discovering Our World*, and is very scarce. "About half the tricks in this booklet [are from the books and] the remaining are from my personal collection of more than 500 science tricks gathered from all parts of the world."

2. *Magic for the Science Class* (Scott, Foresman, and Company, [1941]).

   A similar booklet for the series *Science Problems*.

   * "Magic for the Science Class." In *Impromptu*, edited by Todd Karr (Miracle Factory, 2015) 747-751.

19.3 Book Introductions

1. Bill Simon, *Card Magic for Amateurs and Professionals* (Dover, 1998) viii-ix.
2. Paul Curry, *Magician's Magic* (Dover, 2003) 1-4.
3. John Mulholland, *Beware Familiar Spirits* (Scribners, 1979) [1-5].
4. S. W. Erdnase, *The Expert at the Card Table: The Classic Treatise on Card Manipulation* (Dover, 1995) vii-ix.
5. Diamond Jim Tyler, *Bar Bets to Win Big Bucks* (Dover, 2020) ix.

   Easy-to-Master Mental Magic by James Clark (Dover 2010) was advertised as having a Martin Gardner introduction. but he died before it could be included.

19.4 Articles

1. "Tricks with Bottles." *Mechanix Illustrated* (June 1938) 99, 134, 144.

   * "Tricks with Bottles." In *Martin Gardner Presents* (Richard Kaufman and Alan Greenberg, 1993). after appendix.
   * "Tricks with Bottles." In *Impromptu*, edited by Todd Karr (Miracle Factory, 2015) 752.

2. "Four Simple Tricks." *Mechanix Illustrated* (December 1939) 112, 145.

3. "Eggs-periments." *Sunday Times* [Chicago] (March 24, 1940) 16M-17M.

  - "Eggs-periments." In *Martin Gardner Presents* (Richard Kaufman and Alan Greenberg, 1993). after appendix.
  - "Eggs-periments." In *Impromptu*, edited by Todd Karr (Miracle Factory, 2015) 751. Reduced and hard to read.

4. "Try These on Your Friends." *Popular Mechanics* (December 1941) 78-79.

  - "Try These on Your Friends." In *Magic for Everybody* (Popular Mechanics Press, 1944) [15-16]. Number 29 in the "Little Library of Useful Information"
  - "[Selected tricks]." In *The Boy Magician* (Hearst Books / Sterling, 2008). Several separated pages, with photos (of Gardner) on pages 96 and 118

  A collection of magic all drawn from old issues of *Popular Mechanics*.

5. "Kup's Column." *Chicago Times* (May 28, 1947) 47.

  This is a regular feature of columnist Irv Kupcinet. Gardner was the publicist for the annual S.A.M. conference held in Chicago that year, and this column uses the text of his press release.

6. "It Happened Even to Houdini." *Argosy* **331**, 4 (October 1950) 52-53.

  Additional material cut from this article appeared in *The Phoenix* (November 17, 1950); see also *Magic* (June 1997).

  - "It Happened Even to Houdini." *Check* **1**, 1 (March 1951) 72-75.
  - "It Happened Even to Houdini." *Adventure: The Man's Magazine of Exciting Fiction and Fact* **140**, 3 (February 1964) 18-19.

7. George Groth [Martin Gardner], "He Writes with Your Hand." *Fate* **5**, 7 (October 1952) 39-43.

  An article to help the career of mentalist friend Stanley Jaks.

8. "Manhattan Magic." *New York Tribune: This Week* (October 5, 1952) 13, 49.

  - *Martin Gardner Presents* (Richard Kaufman and Alan Greenberg, 1993). After appendix.

9. Martin Gardner and John Conrad, "The Murdering Cardshark." *True: The Man's Magazine* **39**, 248 (1958) 40-43, 82-85.

   This lurid account of Erdnase appeared under the heading of "True Crime." Conrad was asked by the magazine to spice it up.

10. "How to Create Two Big Effects–and Two Little Ones." *Time* (July 22, 1974) 59.

    Gardner contributed a number trick and card trick to this sidebar for an article on magic.

11. "A Pyramid Scheme that Works." *MAA Online* (November 6, 2006). Effect written up by Colm Mulcahy to accompany an interview. www.maa.org/columns/colm/cardcol200610.html

19.5   Book Reviews

1. *New York Times Book Review* (August 20, 1972) 8.

   John Fisher's Magic Book, John Fisher
   So You Want to Be a Magician, L.B. White

2. *Library of Science* (ca. 1973). Advertising brochure.

   The Illustrated History of Magic, Milbourne Christopher

19.6   Television Commercials

1. *Mini-Magic* (Gerald A Bartell and Associates, [ca. 1952]).

   A series of short films in two parts (the effect and the method) meant to be used to bracket any normal television commercial; the viewer would stayed tuned through the commercial to find out how the trick was done.

2. *Minitrix* (Gerald A Bartell and Associates, [ca. 1952]).

   A similar series of short films.

19.7   Promotional Puzzles and Effects

1. [Promotional Cards] (Kirby-Cogeshall-Steinau Co., 1939-1940).

   Magic Tap-a-Drink Card
   What is Your Favorite Dessert?
   Square and Circle Puzzle
   Double Circle Puzzle
   Chinese Ho Ho Game
   Lightning Picture Change
   (Latter idea described in Fulves' *Self-Working Paper Magic*, pp. 137-8)

   - *Martin Gardner Presents* (Richard Kaufman and Alan Greenberg, 1993) 336-344.

2. *Mysterious Question Card* (unknown).

   - "Mysterious Question Card." In *Mathematics, Magic and Mystery* (Dover, 1956) 107.

3. [Premium cards] (Post Cereals, ca. 1955).

   The Magic Colors Trick
   The Magic Spelling Trick
   The Checkerboard Trick
   The Magic Window Trick
   The Laughing Elephant

   - *Martin Gardner Presents* (Richard Kaufman and Alan Greenberg, 1993) 354-370.

Nordisk Matematisk Tidskrift

Årgång 57   Nr 1   2009

NCM

Chapter 20

Mathematical Tricks

It is hard to draw the line between Gardner's "magic" and "mathematical recreations." Indeed, one of the secrets of his remarkable success in popularizing mathematics was to draw on the mystifying and entertaining aspects of mathematical principles. A "puzzle" in one magazine could often be rewritten as a "trick." I have included in this chapter those citations which are in this borderland. Many of these first appeared in the journal *Scripta Mathematica*.

1. *Mathematics, Magic and Mystery* (Dover, 1956).

   Illustrated by the author.
   A force book; the 15th word of each chapter is "of."

2. *Mental Magic: Surefire Tricks to Amaze Your Friends* (Sterling, 1999).

   - *MENSA Brain Twisters* (Main Street / Sterling, 2004).

     Packaged with similar books by Karen C. Richards and Helene Hovanec. No reason for MENSA label.

   - *Mental Magic: Surefire Tricks to Amaze Your Friends* (Dover, 2010).
   - *Mental Magic: Surefire Tricks to Amaze Your Friends* (Goodwill Publishing House, 2012).

245

20.1   Book Introductions

1. William Simon, *Mathematical Magic* (Dover, 1993) 5-6.
2. S. Brent Morris, *Magic Tricks, Card Shuffling and Dynamic Computer Memories* (Mathematical Association of America, 1998) xvii-xviii.

20.2   Articles

1. "Mathematical Card Tricks." *Scripta Mathematica* **14**, 2 (1948) 99-111.
2. "Mathematical Tricks with Common Objects." *Scripta Mathematica* **15** (1949) 17-26.
3. "Topology and Magic." *Scripta Mathematica* **17** (1951) 75-83.
4. "Mathematical Tricks with Special Equipment." *Scripta Mathematica* **19** (1952) 237-249.
5. "Paper Folding." *Encyclopedia Britannica: 14th Edition* **17** (1968) 279.
6. "The Magic Calculator." *New York Times Magazine* (January 18, 1976) 71.
7. "Office of the Occult." *Games* **9**, 1 (January 1985) 12, 60.

   - "La Oficina de lo Oculto." *Cacumen* (March 1985) 44.
   - "Office of the Occult." *Games* (October 2009) 4, 58.

8. "Magic of a Sort." *REC Newsletter* (January-March 1989) 8-10.
9. "Digital Prestidigitation – Steps toward a computer magic package." *Algorithm* **1**, 2 (January/February 1990) 2-3.
10. "Calculator Tricks to Amuse and Bemuse: Variations on the 12345679 Trick." *Recreational & Educational Computing* [REC Newsletter] **6**, 3 (May 1991) 10.

    - *GW* (2001) 83-84.

11. "Puzzles." *Ka'na'ta: Canadian Pacific Airlines Magazine* (Fall 1980) 76, 78.

    Taken from *Encyclopedia of Impromptu Magic*

12. "Mirror Magic." *Zigzag* 1 (April/May 1995) 32-33, 43.
13. "Presti-digit-ation." *Zigzag* 2 (June/July 1995) 44.
14. "Calculator Craziness." *Zigzag* 3 (September 1995) 22-23, 42.

    - *GW* (2001) 85-89.

15. "Try This Trick." *Recreational & Educational Computing* [REC Newsletter] **11**, 1 (July/August 1996) 6.

    Letter explains the Steinmeyer nine-card trick

16. "A New-Year Curiosity for 1997." *Recreational & Educational Computing* [REC Newsletter] (1996) 6.
17. "A New-Year Curiosity for 1998." *Games* (February. 1998) 8.
18. "Mysterious Precognitions." *Word Ways* **31**, 3 (August 1998) 175-177.
19. "Ten Amazing Mathematical Tricks." *Math Horizons* **6**, 1 (September 1998) 13-15, 26.

   • *GW* (2001) 235-240.

20. "Ten Mathemagical Tricks." *Games* (May 1999) 50-51, 44.
21. "Curious Counts." *Math Horizons* **10**, 3 (February 2003) 20-22,26.
22. "Toilet Paper Tube Tricks." *Games* (February 2007) 80.
23. Donald E. Simanek, "Martin Gardner's Mathemagic." *Make* **13** (2009) 174-174.

   Simanek presents some magic (e.g. the Gilbreath principle) tricks Gardner showed him.

24. "An Amazing Mathematical Trick with Cards." *Normat: Nordisk Matematisk Tidskrift* **57**, 1 (2009) 32-33.
25. "An Amazing Spelling Trick." *Word Ways* **42**, 1 (February 2009) 68.
26. "Magic with Bizarre Poems." *Word Ways* **42**, 2 (May 2009) 125.
27. "Word Magic." *Word Ways* **42**, 3 (August 2009) 202-206.
28. "An Amazing Mathematical Card Trick." *Games* (February 2010) 4.
29. "Word Ways Magic." *Word Ways* **43**, 2 (May 2010) 93 .

   • "Word Ways Magic." *Word Ways* **43**, 3 (August 2010) 166. (Tribute reprint.)

30. "More Martin Magic." *Word Ways* **46**, 2 (May 2013) 92.

   Reproduction of 1998 letter to Ross Eckler

20.3   Published Correspondence

1. *Life* **10**, 9 (March 3, 1941) 8.

   Mentions "Match-ic" and the Nazi Cross joke; response to an article with match tricks (February 10, 1941, pp. 12-14.)

2. *Word Ways* **31** (November 1998) 273.

   A very short addendum to "Mysterious Precognitions"

Martin Gardner, *Column Editor*
Hendersonville, NC 28792

Mysterious Siphon

Using a Bunsen burner to heat a glass tube, bend the tube into the shape shown. If you plunge this into a container of water, as soon as bend A is below the water's level, the tube will become a siphon that will drain off all the water from the container! It is the simplest of many different self-starting siphons that have been invented, some going back to ancient Greece.

Why does it work? When the water rushes through the tube to seek its own level, its inertia is sufficient to carry it above bend B and start the siphon working.

Chapter 21

Physics Tricks

Martin Gardner has always had an interest in tricks that had a scientific basis. Perhaps this is related to his mysterian philosophy, that the world around us is mystifying. In any event he clearly collected such effects from an early age. Before the war, he had assembled two booklets for teachers who used the textbook series *Discovering Our World*: *Magic for the Elementary Science Class* (Scott, Foresman, and Company, [1941]) and *Magic for the Science Class* (Scott, Foresman, and Company, [1941]). The introduction stated "About half the tricks in this booklet [are from the books and] the remaining are from my personal collection of more than 500 science tricks gathered from all parts of the world" but Gardner recalled this as a hyperbolic remark by the editors.

Many magic effects use the properties of physical objects and these are scattered throughout Gardner's magic corpus. However they are most concentrated in his *Encylopedia of Impromptu Magic* (Magic Inc., 1978). At the end of his tenure at Parents Institute, Gardner put out a book designed to introduce physics tricks to children: *Science Puzzlers* (Viking Press, 1960), later retitled *Entertaining Science Experiments with Everyday Objects* (Dover, 1981).

21.1   Columns

Beginning in 1990, Gardner began a long-running column, "Physics Trick of the Month," in *Physics Teacher* that appeared in most, but not all, issues. It ended in October 2002. Beginning in June 1994, these were reprinted in *Magic* magazine, appearing irregularly as "Martin Gardner's Corner." *Magic* reprinted them in a very different order, omitting those that were of no interest to magicians, and including some contributions by Gardner that had not appeared in *Physics Teacher*. *Magic* used new illustrations by Tom Jorgenson, replacing the crude drawings in *Physics Teacher*.

These were later reprinted, using the Jorgenson illustrations, as *Science Magic* (Sterling, 1997), later retitled *Science Tricks* (Sterling, 1998). This book is designated below as SM. A second book of the remaining tricks was issued as *Smart Science Tricks* (Sterling, 2004), using new illustration by Bob Steimle. This book is designated below as SST. There were additional tricks in both books, not appearing in either periodical. The following table attempts to make sense of all this.

There are some tricks from *Physics Teacher* that had appeared earlier in the book *Science Puzzlers*. In particular the column from September 1992 appeared as "Circles on the Card" (p. 30); January 1997 appeared as "Blowing out the Candle" (p. 104) and "Dancing Dime" (p. 102); May 2000 appeared as "Watermark Writing" (p. 74); May 2001 appeared as "Magic Picture" (p. 69). There are a couple of other similarities but the *Physics Teacher* column was almost entirely new material. (Note that September 1992 was omitted from SM.) Further *Smart Science Tricks* contains material drawn from *Science Puzzlers*: trick 1 appeared on p. 72, trick 15 on p. 116, trick 16 on p. 20, trick 23 on p. 95, and trick 28 on p. 91.

	Physics		*Magic*		Book	
Title	Date	Pg.	Date	Pg.		Pg.
Falling Keys	9/90	390	6/94	59	SM	34
Cartesian Matches	10/90	478	8/96	85	SM	7
A Square that	11/90	562	2/96	73	SM	70
Ain't There						
Transporting an Olive	1/91	51	5/96	73	SM	44
Roly-Poly Folder	2/91	107	11/96	79	SM	35
Stabbing an Eggshell	3/91	149	11/95	78	SM	36

Title	Physics Date	Pg.	Magic Date	Pg.		Book Pg.
Where Does the Water Go	4/91	241	5/94	68	SM	71
Waltzing Eggshell	5/91	315	11/94	69	SM	37
Frustrating Papers	9/91	416	10/95	83	SM	45
Balancing a Book	10/91	460	1/97	77	SM	48
Dancing Triangle	11/91	538	8/94	67	SM	14
Enchanted Die	12/91	587	8/95	71	SM	72
A Mirror Paradox	1/92	-			SM	68
Invisible Glue	2/92	85	9/94	64	SM	18
Biased Penny	3/92	158	9/96	79	SM	30
April Bills	4/92	215			SM	88
Left- or Right-Eyed	5/92	292	7/94	61	SM	73
Mysterious Circles	9/92	374	7/95	73	SST	86
Climbing Bear	10/92	422	6/99	75	SM	49
Puzzling Quarters	11/92	501	10/96	89	SM	38
Bent Playing Card	12/92	532	10/94	67	SM	74
Zombie Glass	1/93	46	4/96	87	SM	75
Blacker than Black	2/93	94	6/95	69	SM	69
Gorilla Effect	3/93	167			SM	8
Somersault Shell	4/93	204	6/96	71	SM	9
Bernoulli's Principle†	5/93	304	5/95	61	SM	20
Candle See-Saw	9/93	382	4/95	88	SM	23
An Illusion of Weight	10/93	409	7/96	73	SM	76
Two Corking Good Examples	11/93	477	3/95	66	SM	13
Colored Shadows	12/93	-	12/94	65	SM	63
Curious Feedback	1/94	59	2/95	67	SM	50
Bottle, Hoop, and Dime	2/94	80			SM	39
Rotating Egg	3/94	189	1/95	65	SM	40
Magnetized Pencils	4/94	237	2/97	73	SM	77
A Puzzling Moo Horn	5/94	314			SM	58
Rising Marble	9/94	381	9/97	75	SM	41
A Mirror Paradox	10/94	404	3/97	75	SM	64
Miniature Rocket	11/94	488			SM	24
A Snap and Drop	12/94	537			SM	29
Two Ten-cent Betchas	1/95	51	6/97	76	SM	10
Pulfrich Illusion	2/95	117	10/98	75	SM	65
A Water Transfer	3/95	146	8/97	81	SM	11
Pool Hustler Scam	4/95	220	3/96	73	SM	42
Water Level Riddle	5/95	274	11/97	81	SM	12

Title	Physics Date	Pg.	Magic Date	Pg.	Book	Pg.
Retinal Retention	9/95	366	10/97	85	SM	66
Which Thread?	10/95	478	4/97	79	SM	43
Sneaky Switches	11/95	515	5/97	73	SM	55
Three for Bernoulli	2/96	79			SM	19
Marble and the Cork	3/96	146			SM	15
Unbreakable Balloon	4/96	233			SM	22
A Blacklight Code	5/96	318			SM	67
Ball that Rolls Uphill	10/96	461	1/98	73	SST	134
How Far Apart	11/96	486	3/98	79	SST	85
Unreversed Reflection	12/96	563	4/98	79	SST	82
Two Stunts with a Bottle	1/97	53	6/98	72		
Two Shades of Gray	2/97	119			SST	94
Crazy Bounce	3/97	159			SST	37
Amazing Gender Indicator	4/97	239			SST	97
Zero Gravity	3/98	184	5/99	75	SST	10
Tabletops Illusion	4/98	252	3/99	81	SST	96
An Apparition	5/98	317	2/99	87	SST	80
Three Tennis Balls‡	9/98	343	4/99	76		
Catch the Dice	10/98	437	1/99	84	SST	38
Revolving Circle	12/98	543	12/98	86	SST	84
Archimedes' Pump	1/99	41			SST	12
Three Switches Puzzle	2/99	88			SST	110
Jumping Pencil	3/99	178	8/99	83	SST	64
Two Stunts with Eggs	4/99	245	10/99	85	SST	18
Reverse Jack's Profile	5/99	318	9/99	79	SST	8
Match Penetration	9/99	382	5/00	81	SST	95
Two Drop Tasks††	10/99	76	12/99	92	SST	42
Mirror Levitations	11/99	468			SST	81
Two Rollers and a Yardstick	12/99	546	4/00	77	SST	47
A Pebble Curiosity	1/00	40			SST	15
A Slinky Problem	2/00	78	8/01	82		
Knock off the Match	3/00	191	2/02	76	SST	129
Rotating Cylinders	4/00	254			SST	30
A Watermark Code	5/00	301	4/02	84	SST	23
A Pinhole Paradox	9/00	372	7/01	73		
Rotating Face	10/00	425	11/01	71	SST	92
Electric Grape	11/00	492	5/01	82		

	Physics		*Magic*		Book	
Title	Date	Pg.	Date	Pg.		Pg.
Broken Symmetry	12/00	564	9/01	118	SST	108
Water Down the String	1/01	36	3/03	94	SST	16
Water Knot	2/01	107	10/01	75	SST	13
Shuffled Pages	3/01	178	6/01	83	SST	41
Suspended Bottle	4/01	254	8/03	100	SST	39
Magic Jigger	5/01	302	5/02	84	SST	21
Estimating Height	9/01	370	9/03	110		
Unbroken Tissue	10/01	435	11/03	119	SST	35
Bronx Cheer Effect	11/01	490	10/03	111	SST	87
Mysterious Siphon†‡	12/01	561	1/03	102	SST	22
Twirled Ring	1/02	51	6/03	110	SST	33
A Spooky Rotation	2/02	123	10/02	94	SST	102
Floating Vase	3/02	187			SST	99
A Playing-Card Soliton	4/02	229				
A Newspaper Trick	5/02	313			SST	34
More or Less than Half Full?	9/02	378				
A Paper Pistol‡‡	10/02	442	4/04	102	SST	120
Light from the Wrong End			12/95	83	SM	25
Mysterious Balloon			1/96	83	SM	20
Penny for Your Thoughts			7/97	81	SM	26
Rotating an Object withYour Mind			12/97	75	SST	136
Psychic Motor			9/98	74	SM	27
Swinging Cups			11/98	83	SM	46
Impromptu Haunted Napkin			9/00	74		
Levitated Paper Clip			6/04	103	SM	51
How to Measure Volume					SM	16
Three Jets					SM	17
Twisty Snake					SM	28
Balancing Silverware					SM	31
Make a Magic Bird					SM	32
Floating Paper Cup					SM	52
Psychokinesis?					SM	53
Pill Bottle and Paper Clip					SM	54
Electric Pickle					SM	56-7

Title	Physics		Magic		Book	Pg.
Perpetual Motion					SST	74
Whirling Crackers and Pinwheels					SST	76
Nested Glasses or Cups					SST	78
Vanishing Coin					SST	88
Make a Kaleidoscope					SST	89
Two Ways to See Your† Thumbprin					SST	90
Flexible Ovals					SST	100
Vanishing Smuddge					SST	103
3-D Optical Illusion					SST	104
Funny Pencils					SST	105
Ghostly Circles					SST	106
Two Squares					SST	111
Fun with a Mobius Strip					SST	112
Turning a Triangle Upside Down					SST	114
Knot a Pentagram					SST	115
Paper Cup Telephone					SST	116
Paper Cup Squawker					SST	118
Vibrating Card					SST	119
Stop Your Pulse!					SST	122
Find the Penny					SST	123
It's Not Easy					SST	124
Which is Which?					SST	126
Suspended Pencil					SST	127
A Crawling Paper Clip					SST	130
Retrieve the Paper Clip					SST	132
Paper Clip Chain					SST	133
A Peculiar Balance					SST	137
Rising Hourglass					SST	138
A Balancing Feat					SST	140

† title in *Magic* is "Headline"

‡ title in *Magic* is "Three Balls High"

†† title in *Magic* is "Three Drop Tasks"

†‡ title in *Magic* is "Mysterious Glass Straw"

‡‡ title in SST is "A Paper Popper"

21.2 Published Correspondence

1. "Coat Hanger," *Physics Teacher*, vol. 43, January 2005, p. 4.

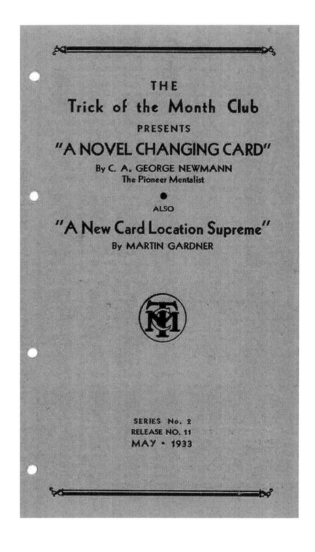

Chapter **22**

Magic Literature

This chapter includes those citations that would normally be only available to magicians. Magic publishers typically have small print runs and so most of these items quickly end up in the hands of a small number of collectors and go off the market. In recent years electronic facsimile collections are becoming more common.

Unfortunately, the amount of detail in the citations below varies a great deal. This is a reflection of the variety of sources I have used that were often second-hand. In fact, most of the magic periodicals and magic anthologies have not been inspected and verified first-hand.

All citations from magic periodicals are included here. Many of these citations are problematic. Often a trick of Gardner's is in fact written up by the editor, and occasionally Gardner will write the text of an effect created by others. It is left up to the editor to decide if a trick (or handling) deserves a full treatment or a short mention in an editorial column. In addition, some citations represent new ideas, while others are small incremental improvements. No serious effort has been made to distinguish these.

22.1 Books

1. *Encylopedia of Impromptu Magic* (Chicago: Magic Inc., 1978).

   Collected from a column that ran in *Hugard's Magic Monthly* from March 1951 until March 1958.

2. *Martin Gardner Presents* (Kaufman and Greenberg, 1993).

   Edited by Matthew Field, Mark Phillips, Harvey Rosenthal, and Max Maven. This volume collects virtually all of Gardner's uncollected tricks and effects from over 60 years. Contains an introduction by Max Maven, an appreciation by Stephen Minch, and a bibliography by Dana Richards.

3. *Impromptu* (Chicago: Miracle Factory with Magic Inc., 2015).

   An expanded edition of *Encylopedia of Impromptu Magic* edited by Todd Karr. Dust-jacket designed by [Jim] Steranko. Foreword by Teller. Preface by John Fisher. "A Brief History of Impromptu." by Todd Karr. Interview by Don Albers from 2005 *College Mathematics Journal*, Gardner's original introduction. "Impromptu Interview," with Todd Karr (by telephone 2005). "The Legendary Files," by Mark Setteducati. "Notes on the Notes," by Todd Karr. Photocopies of unused file cards with introduction, 175 pages. Forty pages of "Letters from Martin." Reprints of four booklets: *Math-ic. After the Dessert, Magic for the Science Class, Over the Coffee Cups.* Sixty pages of "Gardner's Last Revisions." Group of photos. Index.

22.2 Pamphlets

1. *Match-ic: More Than 70 Impromptu Tricks with Matches* (Ireland Magic, 1935).

   This booklet, more than his others, was a compilation of tricks from his files.

   Wooden Matches: magnetized matched, escapo, striking en passant, match box monte, natch and handkerchief, another method, still another method, and still another, electric match, invisible hair, blowing through the sleeves, vanishing penny, fly apart matches, odd and even, tramps and geese, which hand?, penetrating matches, three heaps, three match heads, musical fork, acrobatic box, matches through the table, mysterious mark, contrary match box

Paper Matches: light from the wrong end, smoke from nowhere, folder and string, reappearing matches, divination, parlor mind reading

Trick, Jokes and Catches: blowing through a tumbler, iron match, lighted match from box. smash it, match munition, no light, humming match, strength tester, burned fingers, walking match, vanishing matches, pop up, match head submarines, prolonged kiss, light for two, roly poly box, kiddie kar, trick cigar, rising match box, automatic light, oversize, striking safety matches without the box, cover change, chin penetration

Puzzles: perfect square, equation, star, what happens, drop it, crosses, match - tumblers - salt shaker, linked hands, puzzle lift, puzzle split, three match puzzle, easy when you know.

- *Match Magic: More Than Seventy Impromptu Tricks with Matches* (Piccadilly Books, 1998).

  Tricks are reordered, with 7 tricks omitted and 3 new puzzles.

- *Match-ic* (Brownstone Books / Wildesidebooks, n.d.).

  Facsimile of original. Pirate edition.

- "Match-ic." In *Impromptu*, edited by Todd Karr (Miracle Factory, 2015) 733-740.

2. Joe Berg, *Here's New Magic* (Joe Berg, 1937).

This was ghosted by Gardner.

3. *12 Tricks with a Borrowed Deck* (L. L. Ireland, 1940).

A variant edition can be identified by the exclusion of a Figure 1 on page 5. Illustrated by Tarbell. Introduction by Paul Rosini.

Improved Topsy-Turvy Deck
Face to Face Fantasy
Double Climax Speller
Never-Miss Stop Effect
Big Casino Count Down
Six off Spades Count Down
Two Piles and Subtract
Do As I Do
Eye-Pop Routine
Sympathetic Cards

Vanishing and Reappearing Card
Improved Lie Speller

- *12 Tricks with a Borrowed Deck* (Magic Inc, 1964).

4. *After the Dessert* (1940).

The first edition is just a mimeographed version illustrated by the author.

The Curious Fold
Naming the Date on a Borrowed Bill
Four Piles and a Dollar Bill
The Unreversed Word
The Magic Inhale
Biting the Cigarette
Coin Through the Plate
Vanishing Coin
Testing the Half Dollar
Lighting a Match Twice
Color Changing Heads
Folder Mathematics
Shooting the Match
Match Folder Wager
The Nazi Cross
Tapping Table Objects
Japanese Paper Bird
Vanishing Coffee Steam
Traveling Fountain Pen
Table Knife Through Body
Breaking the Spoon
Bending the Spoon
Swallowing the Knife
Musical Knife
Spoon to Knife
Vanishing Spoon
Character Reading from the Teeth
Improvised Brassiere
The Invisible Hair
Vanishing Salt Shaker

- *After the Dessert* (Max Holden Magic, 1941).
- *After the Dessert: Second Edition* (London: L. Davenport & Co, 1942). Davenport printed under Max Holden on title page

- "[After the Dessert]." In *The Jumbo Book of Magic*, edited by Magic Magazine Editorial Board (Drake Publishers, 1976).

  Uncredited facsimile on pages 13-26. (Discussed in *Genii* **40** (1976) p. 738)

- *After the Dessert* (D. Robbins & Co., 1983).
- "After the Dessert." In *Impromptu*, edited by Todd Karr (Miracle Factory, 2015) 741-746.

5. *Cut the Cards* (Max Holden Magic, 1942).

A Close Fit
The Unconfused Joker
Double Vanish and Recovery
Naming the Card Cut
Use Your Head
The Surprised Gambler
Paper Clip Discovery
Vanish and Spell
The Swizzle Stick Cipher
A New False Cut
A Miracle - Maybe
A Face to Face Routine
Tack it
For the Man in Uniform
A Curious Card Vanish
X-Ray Touch

- *Martin Gardner Presents* (Richard Kaufman and Alan Greenberg, 1993) 379-397.

6. *Over the Coffee Cups* (Montandon Magic, 1949).

Sugar Cubes: cube levitation, steady hands, five cube pick-up, mysterious initials, sugar penetration, hand to hand, cubes - paper - coins

Folding Money: tender and private, striking match on bill, vanishing creamer, Berland's vanishing bill, mushroom - key - states, valuable paper wad, indestructible bill, paper to bill

Silverware: balancing knife and spoon, catching the salt, cross of knives, rotating knife, diver in the spoon

Paper Matches: floating match, grasshopper matches, blowing out the match, happy birthday, have a light, kiddie kar, rocket matches, match penetration, match penetration - 2, Mike and Ike, the three bears, Doc Zola's magnet routine, the hobo's prayer, match head amputation; Napkins: phantom napkin, spooky monkey business, saucy old lady, stretching the napkin, lemon bug, pull apart napkins

Cigars: rising cigar wrapper, jumping cigar band, Indian gifts cigar antics

Coins: sixteen coin puzzle, three stunts with a dime, spinning the half dollar, sucker bet, penny puzzle, try this on your ham sandwich

Glasses: I. Q. test for waitresses, brim to brim, problem of the six glasses, the pyramid, penny game, cross of pepper, match - nickel - glass, glass and check

Cigarettes: fireproof fingers, ear inhale, cigarette pack gags, jugglery, smoke rings, vanishing smoke, jumping and rising cigarette, buck teeth

Miscellaneous: handkerchief escape, stirring stick movies, catching the check, swizzle stick gags, knuckle poppers, the fifty sponge balls, up the scale, financial spectacles, mental prediction, crazy crackers, Schoke's wise crackers, thumb tip quickie, sailor and the doughnut, ring vanish, between the teeth.

- "Over the Coffee Cups." In *Impromptu*, edited by Todd Karr (Miracle Factory, 2015) 753-763.

7. Uriah Fuller [Martin Gardner], *Confessions of a Psychic* (Karl Fulves, 1975).
8. Uriah Fuller [Martin Gardner], *Further Confessions of a Psychic* (Karl Fulves, 1980).
9. *A Die of Another Color* (Karl Fulves, 1995).
10. *The Gardner-Smith Correspondence* (H&R Magic Books, 1999).

The letters between Gardner and Erdnase's illustrator Marshall D. Smith. Preface by Gardner, with "a few additional facts" by Richard Hatch; Charlie Randall also listed under copyright. Limited to 250+26 copies.

11. Dr. Milton A. Ray [Martin Gardner], *Ain't That a Peach?* (Phondlehiene Gootch, 2004).

"A rude, crude and lewd selection of stunt, schtick and trick for the undiscriminating entertainer." Introduction by Armand T. Ringer [Martin Gardner]. The introduction indicates Ray is a Pentecostal pastor, hence "Dr." Ray is "D.D."; the explanation for all this is the fact "a dirty old man" is an anagram of "Milton A. Ray, DD." Illustrated by R. D. Jorgenson. Even the publisher chose to be anonymous!

22.3 Book Introductions

1. Bob Hummer, *Six More Hummers* (L. L. Ireland, 1941) 2.
2. Bill Simon, *Effective Card Magic* (Louis Tannen, Inc., 1952) viii-ix.
3. Paul Curry, *Magician's Magic* (Franklin Watts, 1965) 1-4.
4. Bob Hummer, *Collected Secrets (Second Edition)* (Karl Fulves, 1980) F-H.
5. Karl Fulves, *Swindle and Cheat: A collection of nontransitive games* (Karl Fulves, 1991) 1.
6. Bart Whaley, *The Man Who was Erdnase* (Jeff Busby Magic Inc., 1991) vii.

While Gardner helped with this book and the title page reads "Bart Whaley with Martin Gardner and Jeff Busby" he was not involved in writing this book, about the life and times and techniques of the famous card hustler. However he did sign the Special Autograph Edition. It won a SAM Special Library Award. (This is similar to the situation in 1990 when Bart Whaley indicated his effort was "with Martin Gardner and Jeff Busby," for his comprehensive *Who's Who in Magic*.)

7. Algonquin McDuff [Rhett Bryson and Dexter Cleveland], *Watch Winder Handbook* (Jester's Press, 1992) 4.
8. Diamond Jim Tyler, *Bamboozlers: Volume 1* (Diamond Jim Productions, 2008) 9.

22.4 Articles

1. "The Mystery of Erdnase." In *Souvenir Program* (Society of American Magicians, 1947) unpaginated.

- "The Mystery of Erdnase." In *The Annotated Erdnase*, edited by Darwin Ortiz (Magical Publications, 1991) 263-265.

2. "The Man Who was Erdnase." In *The Annotated Erdnase*, edited by Darwin Ortiz (Magical Publications, 1991) 260-262.
3. "How to Make Magic Changing Pictures." *Magic* (February 1997) 60-64.
4. "Various Anecdota Magicana." *Magic* (June 1999) 112.

    Material cut from the 1950 *Argosy* article that appeared in the *Phoenix*, no. 216 (November 17, 1950) is reprinted here.

5. Richard Hatch, "Searching for Erdnase." *Magic* (December 1999) 67-73.

    Contains a facsimile of Gardner's 1946 letter to Marshall D. Smith and the follow-up.

6. "The World's Second-Best Spelling Trick." *Gibecière* **5**, 1 (Winter 2010) 49-60.

22.5   Book Reviews

1. *Pallbearer's Review* (December 1973) 688.

    John Fisher's Magic Book, John Fisher
    So You Want to Be a Magician, L.B. White

2. "Light from the Lamp." *Genii* **48**, 9 (September 1984) 693.

    Psychic Paradoxes, John Booth

3. *The New Tops* (March 1990) 7-10.

    Stewart James in Print, Howard Lyons and Allan Slaight

4. "Membership has Benefits." *M.U.M.* (February 2006) 83-84.

    Gibercière. Winter 2005, Conjuring Arts Research Center

5. "Impuzzible, Too." *M.U.M.* (March 2006) 88.

    Further Impuzzibilities, Jim Steinmeyer

6. "Mind Games." *M.U.M.* (May 2006) 85.

    Sleight of Mind, George Ulett

7. "Sum Wizardry." *M.U.M.* (June 2006) 83.

   How to Perform Feats of Mathematical Wizardry, Harry Lorayne

8. "Starlight, Starbright." *M.U.M.* (April 2006) 77-78.

   Roy Benson by Starlight, Levent and Todd Karr

9. "Self-Workers." *M.U.M.* (April 2006) 79-80.

   Automata, R. Shane

10. "You Do the Math." *M.U.M.* (August 2006) 77.

    Secrets of Mental Math, Arthur Benjamin and Michael Shermer

11. "16th Century Wonders with Some Hoaxes, Too." *M.U.M.* (December 2007) 72-73.

    Gibercière. Summer 2007, Conjuring Arts Research Center

12. "Always Cut the Cards." *M.U.M.* (April 2008) 68.

    52 Ways to Cheat at Poker, Allan Zola Kronzek

13. "Dead End." *M.U.M.* (April 2008) 74.

    The Complete Idiot's Guide to Street Magic, Tom Ogden

14. "Between the Lines." *M.U.M.* (November 2008) 63-64.

    This is Not a Book, Robert Neale

22.6   Marketed Effects

1. *A New Card Location Supreme* (Trick of the Month Club, May 1933). Series No. 2, Release No. 11.
2. *Gardner's Unique Principle* (L. L. Ireland, [ca. 1935]).
3. *Martin Gardner's Triple Spell Miracle* (L. L. Ireland, 1939).
4. *Gardner's Passe Passe Sponge Trick* (L. L. Ireland, 1940).
5. *The Goose That Failed* (H. Fishlove and Co., 1940).

   Magnetic novelty item. Also sold as "Ma and Pa" (1949) and changed into the "Amazing Scottie Dogs" in the 1950's.

   - *Martin Gardner Presents* (Richard Kaufman and Alan Greenberg, 1993) 345-347, 360.

6. *Martin Gardner's Mother Goose Mystery* (Berland Press, 1941).

   Contains a booklet entitled "Favorite Mother Goose Rhymes"

   - *Martin Gardner's Mother Goose Mystery* (London: Magic Wand, 1953).

     Newly typeset with addendum by George Armstrong
   - *Martin Gardner Presents* (Richard Kaufman and Alan Greenberg, 1993) 348-359.

7. *Are You Psychic?* (L. L. Ireland, 1941).

   Also appeared in the 1943 Ireland Yearbook (pp. 13-15), and as the "Occult Bartender."

8. *Cherchez la Femme* (Roger Montandon, 1946).

   Racy tetraflexagon.

   - *Martin Gardner Presents* (Richard Kaufman and Alan Greenberg, 1993) 361-363.

9. *Occult Bartender* (Magic Inc., n.d.).

   Sold as two pairs of blank pads, with words on the covers, with instructions. Originally appeared as "Are you Psychic?"

22.7   Tricks in Books

A handful of these (such as those in Dover books) are available to the public but are also listed here for consistency.

1. Will Goldston, *Great Magician's Tricks* (Goldston, 1931).

   Four aces and four kings, p. 200

2. Jean Hugard, *Encyclopedia of Card Tricks* (Louis Tannen, n.d.).

   Gardner's unique principle, pp. 149-151

   - Jean Hugard, *Encyclopedia of Card Tricks* (Max Holden, 1937). also 1940.
   - Jean Hugard, *Encyclopedia of Card Tricks* (London: Faber and Faber, 1961).
   - Jean Hugard, *Encyclopedia of Card Tricks* (Dover, 1974).

3. Joe Berg, *Here's New Magic* ([Joe Berg], 1937).

   Gardner's card speller, p. 3
   Gardner's immovable cigarette, p. 15

4. Ed Marlo, *Let's See the Deck* (L. L. Ireland Co, 1942).

   Gardner-Marlo poker routine, p. 28

5. Ed Marlo, *Deck Deception* (Magic Inc., 1942).

   Casanova card trick, p. 5-6

6. Roger Montandon and Logan Wait, *Not Primigenial* (1942). [self-published]

   Through the center, p. 13

7. [L. L. Ireland, (Ed)], *Ireland's Yearbook for 1943* (1943). L. L. Ireland

   Are you psychic?, p. 13-15

   - [L. L. Ireland, (Ed)], *The Yearbook Reader* (Magic, 1977).
     Are you psychic?, p. 136-137

8. J. R. Crimmin (ed.), *Ted Annemann's Full Deck of Impromptu Card Tricks* (Max Holden, 1943).

   An easy lesson [Do as I do], p. 38
   Draw poker plus, p. 57
   Preposterous, p. 63 [also a Jordan variation]
   Newspell, p. 64-5
   Jordan plus Gardner [Preposterous variation], p. 66
   Card minded (with a red/black overhand shuffle), p. 69

   - J. R. Crimmin (ed.), *Annemann's Card Tricks* (Dover, 1977).

9. J. G. Thompson, J. G. Thompson, *My Best* (Louis Tannnen, [1959]).

   Spread, shuffle, spread, p.89
   Pop through cigarette, p.166
   Biographical sketch, p. 358

   - J. G. Thompson, *My Best* (D. Robbins, 2005).

10. Stewart, James (Compiler), *Abbott's Encyclopedia of Rope Tricks, Volume II* (Abbott's Magic Novelty Co., n.d.).

    Martin Gardner's Magic Knot, p. 34, 37
    Martin Gardner's Chefalo Release, p. 34, 36
    Martin Gardner's Impossible Knot, p. 36-39

11. Neal Elias, *At the Table* (George Snyder, Jr., 1946).

    Vanishing card, p. 3-4

12. Warren Wiersbe, *Mental Cases with Cards* (Ireland, 1946).

    Think-a-Card, p. 18

13. Harlan Tarbell, *The Tarbell Course in Magic, Volume 5* (Louis Tannen, 1948).

    'Pop up' cigarette, pp. 52-53
    Immovable cigarette, pp. 53-55
    Card through handkerchief, pp. 84-86
    Wand through the hat, pp. 106-108
    Brief mention in volume 7.

14. Jean Hugard and Frederick Braue, *The Royal Road to Card Magic* (Harper & Brothers, 1948).

    Poker puzzle, pp. 129-130

    - Jean Hugard and Frederick Braue, *The Royal Road to Card Magic* (Faber & Faber, 1949).
    - Jean Hugard and Frederick Braue, *The Royal Road to Card Magic* (World Publishing, 1951).
    - Jean Hugard and Frederick Braue, *The Royal Road to Card Magic* (Dover, 1999).
    - Jean Hugard and Frederick Braue, *The Royal Road to Card Magic* (Martino Fine Books, 2015).

15. Al Leech, *For Card Men Only* (Magic Inc, 1949).

    Elusive jacks, p. 12-15

16. W. F. (Rufus) Steele, *52 Amazing Card Tricks* (1949). [self-published]

    Five, nine, king, p. 11

17. John Scarne, *Scarne on Card Tricks* (Crown, 1950).

    There are many attributions to Gardner throughout this book. However, many tricks are attributed to him incorrectly, while some of his tricks are credited to others.
    It's a natural, p. 23
    The initials will tell, p. 68+
    Five nine king, p. 176
    Sympathetic cards, p. 214
    Impossible location, p. 249

18. Bill Simon, *Effective Card Magic* (Louis Tannen, Inc., 1952).

    Similar twins, p. 151

    • Bill Simon, *Card Magic for Amateurs and Professionals* (Dover, 1998).

19. Harold Rice, *The Encyclopedia of Silk Magic, Volume 2* (Silk King Studios, 1953).

    Pop up cigarette, pp. 601-2

20. Verne Chesbro, Larry West, *Tricks You Can Count On* (1955).

    All the nonconformists, p.31

21. Bruce Elliott, *The Best in Magic* (Galahad Books / Harper and Row, 1956).

    Hand to hand, p. 112
    Ten card deal [variation], p. 193

22. Lewis Ganson, *The Dai Vernon Book of Magic* (London: H Clarke and Co, 1956).

    Cigar vanish, p. 96

    • Lewis Ganson, *The Dai Vernon Book of Magic* (Supreme Magic Co, 1957).
    • Lewis Ganson, *The Dai Vernon Book of Magic* (Harry Stanley, 1957).
    • Lewis Ganson, *The Dai Vernon Book of Magic* (L&L, 1994).
    • Lewis Ganson, *El Libro de Dai Vernon* (Madrid Paginas, 2005).

23. Denhard Harold, *How to Do Rope Tricks* (Magic Inc, 1957). (1982 edition with additions by Don Tanner and Frances Marshall.)

The Impossible Knot, p. 61

24. [Magic Circle], *Magic of the Television Stars* (London: H. Clarke and Co Ltd, [1957]). (Title on cover "Tricks of the Television Stars.")

A Geometrical Vanish, p. 48

25. Bruce Elliott, *Professional Magic Made Easy* (Harper, 1959).

Texture [Pop up cigarette], p. 3

26. Frances Marshall, *The Sponge Book* (Ireland Magic, 1960).

Passe trick, pp. 29-32

27. Harry Lorayne, *Close-Up Card Magic* (Louis Tannen, 1962).

Lorayne's Poker Deal, pp. 153-157

28. Gene Anderson, Frances Marshall, *Newspaper Magic* (Magic Inc, 1968).

Martin Gardner on Afghan Bands, pp. 120-122

29. Lewis Ganson, *Give a Magician Enough Rope ... and He'll Do a Trick* (Harry Stanley, 1968).

Variations on the Hunter knot, p. 68-70

- Lewis Ganson, *Give a Magician Enough Rope ... and He'll Do a Trick* (Supreme Magic Co., 1978).

30. Verne Chesbro, *Ultimate Color Separation No 2* (1969).

unknown

31. Philip R. Willmarth, *Fun with a Handkerchief* (Magic Inc, 1969).

The ghost, p. 5

32. Karl Fulves, *Tricks with Dice* (Karl Fulves, 1970).

The four color problem, p. 25

33. Harold Dennard, *How to Do Rope Tricks* (Magic Inc, 1973).

Impossible knot

34. Karl Fulves, *Riffle Shuffle Set-Ups* (Karl Fulves, 1973).

    Color thot, p. 81
    A betting swindle, p. 88

35. Bob Read, *Thanks to Pepys...* (Read, 1973).

    Brush gag, p. 27

36. Paul Curry, *Never in a Lifetime* (1975). [self-published]

    unknown

37. Jerry Mentzer, *The Magic of Paul Harris* (Jerry Mentzer, 1976).

    Twistover, p. 42 (with Tenkai Ishida)

38. Karl Fulves, *Self-Working Card Tricks* (Dover, 1976).

    Plunger glimpse, p. 14
    Ultra-coincidence, p. 44

39. Frank Garcia, *The Encyclopedia of Sponge Ball Magic* (1976).

    Passe trick, p.107

40. Karl Fulves, *The Magic Book* (Little Brown, 1977).

    Turnabout coins, p. 88
    The red-black swindle, p. 168
    How two cut the aces, p. 168

41. Frederick Braue, *Fred Braue on False Deals* (Jeff Busby, 1977).

    Middle deal variation, p. 26

42. Phil Willmarth, *Magic from the Parade* (Phil Willmarth and Assoc, 1979).

    Impromptu Vanisher, Stretching a Handkerchief, Grandma's Needle, Jumping Hanky, Spy-Key

43. Stephen Minch, *Bob Hummer's Collected Secrets* (1980).

    Hummer's Fortune Telling Book, p. 77, 114
    Paper clip discovery, p.118

44. Alfred B. Leech, *Al Leech's Legacy* (Magic Inc, 1980).

    A typographical error, p. 18-20

45. Stephen Minch, *Jacks or Better* (1980).

   Lorayne's poker deal, p. 3

46. Karl Fulves, *Curiouser* (Karl Fulves, 1980).

   Euclid's vanish, p. 11

47. Karl Fulves, *The Children's Magic Kit* (Dover, 1980).

   Hat or rabbit p. 15
   Vanishing rabbit, p. 16

48. John Mendoza, *Don England's TKOs (Second Edition)* (Thinker's Press, 1980).

   Blink change, p. 38

49. Karl Fulves, *Self-Working Table Magic* (Dover, 1981).

   The thirteenth turn, p. 28
   Spell-a-die, p. 29
   Comedy vanish, p. 84
   Looking-glass logic, p.96-98
   Pen Kink, p. 118

50. Karl Fulves, *Octet* (Karl Fulves, 1981).

   Quant quirk, p. 18
   A up/down bet, p. 20
   Rubik Cube swindle, p. 23

51. Karl Fulves, *Side Steal #4: Cards* (Karl Fulves, 1981).

   Daley's card thru hank, pp. 11-12

52. Karl Fulves, *Robert Neale's Trapdoor Card* (Karl Fulves, 1983).

   Twist and link, p. 3

53. Karl Fulves, *Self-Working Number Magic* (Dover, 1983).

   Locked room mystery, trick p. 29
   Crazy calendar, trick p. 73

54. Tom Craven (Ed.), *Pallbearer's Review Lecture* (undated).

   Odd and Even Dice, p. [16]

55. Algonquin McDuff [Rhett Bryson, Jr. and Dexter Cleveland], *Spirit Cloth Book* (Jester's Press, 1984).

   Suggestions, pp. 11-13

56. Karl Fulves, *More Self-Working Card Tricks* (Dover, 1984).

Color of thought, pp. 28-30
Wild Aces, p. 134

57. Phil Willmarth, *Magic from the Parade #1* (1984).

Impromptu vanisher, p. 2
Stretching a handkerchief, p. 3
Grandma's needle, p. 4
Jumping hanky, p. 5

58. Karl Fulves, *Riffle Shuffle Technique: Part III* (Karl Fulves, 1984).

Daley red/black shuffle, p. 116

59. Karl Fulves, *Self-Working Paper Magic* (Dover, 1985).

Paradox papers, pp. 18-20

60. Frederick Braue, Jeff Busby, *The Fred Braue Notebooks, Vol. 4* (Jeff Busby Magic, 1985).

The fancy false cut, p. 3
The unconfused joker, p. 25

61. Frederick Braue, Jeff Busby, *The Fred Braue Notebooks, Vol. 5* (Jeff Busby Magic, 1985).

Naming the card variant, p. 14

62. Phil Goldstein [Max Maven], *ESP Dice Mysteries* (manuscript, 1986).

The Thirteenth Turn, pp. 4-5

63. Karl Fulves, *Contemporary Rope Magic: Part II* (1986).

Christopher's Knot, p. 30

64. Karl Fulves, *Contemporary Rope Magic: Part III* (1986).

Las Vegas Knot Knot, p. 58

65. Karl Fulves, *Self-Working Handkerchief Magic* (Dover, 1988).

The mouse, p. 87

66. Karl Fulves, *Prototype* (1989).

A trio of tricks, p. 19-21

67. Karl Fulves, *Quick Card Tricks* (1989).

   Handling of Rightful Place, p. 21

68. Rhett Bryson, *Small Magical Ideas* (Jester's Press, 1989).

   Dancing pencil, p. 21
   How did I know, p. 22
   Jumping band, p. 23
   Rising pencil, p. 24
   Flapping butterfly, p. 25
   A simple reversal, p.26

69. Karl Fulves, *Self-Working Rope Magic* (Dover, 1990).

   Scissorcut, p. 88

70. *The Klutz Book of Magic* (Klutz Press, 1990).

   This is a spiral bound book, by John Cassidy and Michael Stroud, that contains unspecified contributions by Gardner, who was a "technical consultant."

71. Stephen Minch, *Collected Works of Alex Elmsley, Vol. 1* (L&L Publ, 1991).

   Card hopper, p. 371 (with Elmsley)

72. Karl Fulves, *Swindle & Cheat* (1991).

   Handling of 36 Number Dice, p. 60

73. Frederick Braue, Jeff Busby, *The Fred Braue Notebooks, Vol. 6* (Jeff Busby Magic, 1992).

   Double speller / Surprising speller, p. 22

74. Frederick Braue, Jeff Busby, *The Fred Braue Notebooks, Vol. 8* (Jeff Busby Magic, 1992).

   Double Speller, p. 22

75. Arturo Ascanio, *La Magia de Ascanio* (1992).

   Triple Corte, p. 78

76. Ken Simmons, *Choice Effects, Number One* (1992).

   Vanishing and reappearing card, p. 1

77. T. A. Waters, *Mind, Myth & Magick* (Hermetic Press, 1993).

Psidentify, p. 685

78. Allan Ackerman, *Las Vegas Kardma* (A-1 Multimedia, 1994).

Gardner-Tenkai reverse, p. 102

79. Karl Fulves, *New Card Rises* (1996).

Saliva Card Rise, p. 36-37

80. Paul Harris, Eric Mead, *The Art of Astonishment, Book 1* (A-1 Multimedia, 1996).

Twistover, p. 56 (with Tenkai)

81. Algonquin McDuff [Rhett Bryson, Jr. and Dexter Cleveland], *The Baby Bag* (Jester's Press, 1996).

Ideas: Jumping Bag, Poltergeist Bags, Monte Bags, Two Cents Still, Wandering Nickel, Rubber Ring, A Quart Low, Houdini's Bag, Flying Coins, Super Bag, Pentagon Vision, Mini Color Vision, Zero Sodium Chloride, Passe Passe Marbles, pp. 36-42.

82. G. Livera and J. Racherbaumer, *The Amazing Cigar* (Magic Marketing Concepts, 1997).

Band Dance

83. *Coin Magic* (Klutz Press, 1997).

This little spiral bound book "by the editors of Klutz Press" contains unspecified contributions by Gardner, who is "acknowledged."

84. Algonquin McDuff [Rhett Bryson, Jr. and Dexter Cleveland], *The Money Maker Machine Manual* (Jester's Press, 1998).

Guest Routine , p. 71

85. Karl Fulves, *Calculator Tricks* (1998).

Calculated Swindle, p. 110

86. Roberto Giobbi, *Card College: Volume 3* (Hermetic Press, 1998).

Numerology, pp. 703-702, with Vollmer and Christ

87. Roberto Giobbi, *Card College: Volume 4* (Hermetic Press, 1998).

Stop trick, p. 910, with Horowitz and Giobbi

88. Oliver Erens, *Concertos for Pasteboards* (Hermetic Press, 2000).

The pad spread, p. 144, with Dan Tong

89. Jeff Brown, *Crayon Magic* (SPS, 2000).

Classified Crayons, p. 29
"Pop-Up" Crayon / Color Filter, p. 43-44
Immovable Crayon / Color Sympathy, p. 44-45
"X" Marks the Card, pp. 55-56

90. Jeff Brown, *The Sherlock Holmes Book of Magic* (Picadilly, 2000).

You Can Run But You Can Not Hide, pp. 76-78

91. Karl Fulves, *New Self-Working Card Tricks* (Dover, 2001).

Synchronism, p. 83 (with Tony Bartolotta)
Make the cards match, p. 130 (with Howard Adams)
Mind matrix, p. 138 (with Robert Brethen)

92. Donato Colucci, *Encyclopedia of Egg Magic* (Hermetic Press, 2002).

New method egg balance, p. 109

93. John Carney, *The Book of Secrets* (CarneyMagic, 2002).

Airing the deck, p.64
Wink change variation, pp. 185-186

94. P. Howard Lyons, Max Maven, *Aziz & Beyond* (Hermetic Press, 2002).

PUZ and ZEL, p. 46

95. Steve Beam, *Semi-Automatic Card Tricks: Volume 6* (Trapdoor Productions, 2006).

Airing the deck, pp. 131 (with Andrus)

96. Nick Trost, *Subtle Card Creations: Volume 1* (H & R Magic, 2008).

    The 'one-two-three' trick, pp. 35-37

97. Steve Beam, *Semi-Automatic Card Tricks: Volume 8* (Trapdoor Productions, 2010).

    The 'ace-two-three' trick, pp. 74-76 (with Steve Beam)

98. Roberto Giobbi, *Card College Lightest* (Hermetic Press, 2010).

    Two, six, ten, p. 83

99. Steve Beam, *The Trapdoor, Volume One* (Trapdoor, 2011). Collection of nos. 1-25

    Comment on Upside down, p. 1-3, originally 1983

100. Bob Farmer, *The Bammo Ten Card Dossier* (Magicana, 2015).

    Ten card deal, p. 13-18

101. Jamy Ian Swiss, Johnny Thompson, *The Magic of Johnny Thompson: Volume 2* (Magicana, 2018).

    Tic tac toe force, p. 164-170

22.8   Magic Periodicals

Magic periodicals do not circulate and generally can only be found in the collections of magicians. However in recent decades there has been efforts to digitize these and you can often buy a CD of the entire run of many of periodicals listed below. A lot of the citations were collected in the book *Martin Gardner Presents* (1993).

Abracadabra

"The Oshkosh Trick" **6**, 140 (2 Oct 1948) 158
Fabian's write-up of topological tricks from *SA*. **39** (9 Jan 1965) 87-9.

Almanac

"Hole Second Deal" **2**, 21 (1984) 205.

Apocalypse

"Fibonacci Fantasy" with Stewart James. **1**, 8 (Aug 1978) 88-9.
"All Good Things" with Mitzman and Hovalka. **7**, 5 (1984) 915.

Arcane

Special Issue: Martin Gardner & Friends, 14 (1995).
  "Hand-Held Pasteboard Prevarication Detector," 188-91.
  "Knot from Nowhere & Variations," 191-4. (with Berg)
  "Separating Knots," 196-8. (with Matsuyama)
  "The Pyramid Spell," 198-200.
  "Self-Suspending Matches" 200-2.
  "Quad Rise" with "Tilt Finesse," 202-4. (with Diaconis)
  "Chromo-Clairvoyance" with "Literary Brilliance," 205-8
  "Sefalaljia Sans," 208-9.
  "Two Quick Tricks & Associated Ideas," 210-2. (with Busby)
  "Packet Parity," 212.

The Bat

Two math puzzles in the "Puzzle Corner." 57 (Sept 1948) 418.
(An off-color trick). 59 (Nov 1948) 434.

Chaos

"Racquel Welch and a Tricky Dick." 11 (Jun 1974). (#11 out of order!)
"Think-of-a-four-letter-word Mystery." 34b (Oct 1974).

Charlatan

"Miller's Strip Out." 4 (2000) 43.
"The Giraffe Puzzle." 4 (2000) 48.
"Three Dice Bets." 5 (2001) 62. (Also answer to Giraffe trick.)
"Mate in One." 8 (2001) 107-8.

Cheat Sheet

"Rattle rock spoon." 7 (1991) 69.
"Red-black paradox." 9 (1991) 84.
"Cornered King." 10 (1992) 103.

Chronicles

"Truzzi-Gardner Method." 2 (1978) 1077.
   A levitation behind cloth (with Truzzi)
First Finger Grip. 8 (1978) 1142.
"Two-Coin Vanishes." 8 (1978) 1144.
"Martin Gardner's Ideas (on The Red Prediction)." 8 (1978) 1152-3.
"Psychic Motors" with Fulves. 11 (1978) 1168.
"Two Coin Vanishes." 14 (1979) 1198.
"Boiling Point." 15 (19798 1211-2.
"Gardner's Aces." 20 () 1256. "Four Aces." 24 (1979) 1303-4.
"The Five-Sided Business Card." 30 (1981) 1355-7.
   Also a false snap-turnover.
   Also a comment on a Braue reversal on p. 1362.

The Conjurors' Magazine

"National Conference of S.A.M.." **3**, 3 (May 1947) 12.
   (Publicity write-up of the Chicago convention).
'Where's the Joker." **4**, 11 (Jan 1949) 30.
"Two Six Ten." **4**, 12 (Feb 1949) 10.
"Prediction." **5**, 1 (Mar 1949) 28.
"Was Erdnase Abdul Aziz Khan?." **5**, 6 (Aug 1949) 11-2.
   (With reprint of *Harpers Weekly*, June 26, 1909.)

The Crimp

"How to Change an Apple into a Peach." no. 4.
"Jack Miller's Vanishing Goldfish Bowl." no. 5.

Discoverie

"Short Armed." 1 (1999) 2.
"Pencil Magnet." 3 (2000) 39.
"Thot Card to Pocket." 4 (2000) 43-4.
"Turnaround Knot." 5 (2001) 64-5.
"Coin that Floats." 9 (2004) 120.

Epilogue

"Stock Holdout." 2 (1968) 9.
"More Monge." 4 (Nov 1968) 29.
"Two Chess Wagers." 5 (1970) 34.
"Daley's Card Thru Hank." 11 (Mar 1971) 87.
"The 'Wink' Change." 21 (July 1974) 194.

Epoptica

"Three Ideas Via Martin Gardner:." 4 (June 1983).
    "Rotating Die." 230-1.
    "Chair, Rope, and Finger Ring Puzzle." (Answer pp. 234-5) 231
    "A New Sucker Bet." 233
"Mitsunobu Matsuyama's Chronometer Cards." (Apr 1984) 288-9.
    Yearbook.

The Fine Print

"A New Year Curiosity." 5 (1997) 127.
"Pepper." 6 (1997) 154.
"The 3 Switch Problem." 7 (1998) 203.
More on the 3-switch problem. 8 (1998) 249.
"Five x five." 10 (1999) 345.

The Gen

"A Geometrical Vanish." **10**, 11 (Feb 1955) 319.
"Variations on the Hunter Knot." **17**, 7 (Nov 1961) 184.

Genii

"Match through Match Illusion." **5**, 5 (Jan 1941) 158.
"Improved Dollar Bill Mindreading." **5**, 8 (Apr 1941) 257.
"The Thimble Changes Places." **7**, 1 (Sept 1942) 10.
'The Indestructable Sequence." **54**, 3 (Mar 1991) 305.
"Impromptu Magic for Children." **55**, 9 (Sept 1992) 752-3.
"A Different Sure Thing." **63**, 7 (Jul 2000) 59.
"Tic-tac-toe magic square." **85**, 1 (Jan 2022) 59.

The Hierophant

"The 'Wink' Change." 5/6 (Fall/Spring 1971) 290-1.

Hokus Pokus

"Ein paar Streichholzer." **20**, 2 (1959) 799.
"Ein eigenartigen schnurtrick." **20**, 3 (1959) 855.
"Zundholz-kopfchen." **28**, 1 (1967) 8.
"Lippenstift magie." **28**, 1 (1967) 9.

Hugard's Magic Monthly

"The Flying Cigar Band." **3**, 2 (July 1945) 131.
"Coat Penetration." **6**, 5 (Oct 1948) 475-6.
"Under the Table." **6**, 6 (Nov 1948) 485-6.
    (Weigles's on "Hummer's Card Mystery" and finger mnemonics)
"Scissors Sorcery." **6**, 7 (Dec 1948) 495.
    (Also a science stunt with Xmas lights)
"The Plunger Principle." **6**, 8 (Jan 1949) 505.
"Reversed Subway." **6**, 9 (¿Feb 1949) 513.
"Dollar Bill Folds." **6**, 9 (Feb 1949) 515-6.
    (An "mazuma" fish and other folds)
"Card Suggestions." **6**, 10 (Mar 1949) 525.
    (Some ideas for "Topsy Turvy Deck", a fake center deal,
    "Biddle Move", and "Out of this World")
"Finger Stunts." **6**, 11 (Apr 1949) 535.
    (Variations on pulling off the thumb, flexible hand illusions).
"Flashpaper Pellet Switch." **6**, 12 (May 1949) 545.
    (Stewart Judah's handling of switch and two finger stunts)
"Notes on the I.B.M. Convention." **7**, 1 (June 1949) 555.
    (Observations on Slydini, Schoke, Allerton, Crandall, and Dorny).
"Two Pinkie Tricks." **7**, 2 (July 1949) 565.
"Bert Allerton's Crazy Cards." **7**, 3 (Aug 1949) 575.
"Stunts with Paper" **7**, 4 (Sept 1949) 585-6.
    (Marshall's bunnies, paper movies, Dorny's cardboard snapper)
"Card Suggestions." **7**, 5 (Oct 1949) 595.
    (Jump out card, break control, and Charlier pass location.)
"Comedy Stunts." **7**, 6 (Nov 1949) 605.
    (Gags and tricks using "yakity yak teeth", "endless rope"
    Logan Waits' "pebbles in the shoe.")
"The Afghan Bands"). **7**, 7 (Dec 1949) 615.
    (Brief history and two variations.)

"Card Suggestions." **7**, 8 (Jan 1950) 623.
  (Joe Berg's "Nail Location" and a novel discovery).
"Gum Baffler." **7**, 8 (Jan 1950) 624. (Gumball color divination.)
"The Joe Berg Knot." **7**, 9 (Feb 1950) 630-1.
"Two Cigarette Tricks." **7**, 10 (Mar 1950) 645.
"Card Suggestions." **7**, 11 (Apr 1950) 651.
  (Dr. Daley's "Reversed Card Count" with a variation,
  and variation on Allerton's "Crazy Cards")
"Criss Cross Do As I Do." **7**, 12 (May 1950) 665.
"The Red Black Shuffle." **8**, 1 (June 1950) 674.
"Red Black Riffle Shuffles." **8**, 2 (July 1950) 679, 681.
"More Red Black Riffle Shuffles." **8**, 3 (Aug 1950) 689-90.
"Red Black Riffle Shuffles" (Conclusion). **8**, 4 (Sept 1950) 699, 705.
"Bouncing Putty" (Comedy ideas for Silly Putty). **8**, 5 (Oct 1950) 713.
"The Tenkai Pennies." **8**, 6 (Nov 1950) 721.
"Jack Miller's Ring and Stick." **8**, 7 (Dec 1950) 735.
"Poker Chip Faro Shuffle." **8**, 8 (Jan 1951) 739, 746.
"Eddie Marlo's 'Grab It' ." **8**, 9 (Feb 1951) 751.
"Encyclopedia of Impromptu Magic." **8**, 10 (Mar 1951) 763.
  (Start of monthly series that ran until **15**, 11 (Apr 1958), 131)
"Double Climax." **15**, (Feb 1958) 103.
"Japanese Finger Tie." **15**, 11 (Apr 1958) 123.
"Gypsy Glass Trick." **16**, 10 (Mar 1959) 115.
"Rope and Coat Release." **17**, 3 (Aug 1959) 29.
"Gardner's Card Naming." **18**, 11 (July 1961) 130.
"Precognition with Cards" . **19**, 1 (Sept/Oct 1961) 17.
  (Reprints a card revelation from *SA*; continued to no. 2, Oct 1961)

Ibidem

'The One, Two, Three Trick." 6 (July 1956) 5.
"Principles Only." 6 (July 1956) 9. (An idea based on card centers)
"Flexa Tube Puzzle." 7 (Sept 1956) 13.
  (Flexagon puzzle; answer in no. 9, p. 12)
"Smith's Dilemma." 7 (Sept 1956) 21. (A "blue" number stunt)
"Looped." 10 (June 1957) 9-10. (A topological trick)
Puzzles with cigarettes and matches. 13 (Mar 1958) 5, 15.
"Piano Mover." 16 (Mar 1959) 8. (Piano trick using double lift)
"The Compleat Gardner." 18 (Oct 1959) 3-8.
  (Many ideas, stunts, and tricks.)
"Last August." 20 (May 1960) 16.
  (An idea using a rubber band and finger ring)

"Brain Popper." 23 (Mar 1961) 23. (A prediction paradox)
"All the Nonconformists." 24 (Dec 1961) 29.
"Magical Chairs." 24 (Dec 1961) 30. (A Stanley Jaks prediction)
"Doink." 26 (Sept 1962) 25.
   (A novelty quickie with a hat pin and eraser)
Letter to Chesbro about *Ultimate Color Separation*. 33 (Jul 1968) 20.
Chat about cherry in glass matchstick stunt. 33 (Jul 1968).
"Dam You Charlie Brown." 34/35 (Aug 1969) 39.
   (Dental dam with Lyons' trick)

Interlocutor

(Reprint of a *Washington Post* interview). 19 (1978) 73-4.
(Improvement to a paper trick by Kovachevich). 43 (1980) 170.
(Loaded dice joke). 45 (Mar 1981) 177.

Intermagic

"Von hand zu hand." **2**, 2 (1974) 35.
"Der esoteriche glas-trick." **20**, 2 (1996) 83.
"Eine mutprobe." **21**, 3 (1997) 104.
"Stegreif-hensehen." **21**, 3 (1997) 114.

The Jinx

"Preposterous." 40 (Jan 1938) 268.
"Martin Gardiner's [sic] Manuscript." Winter Extra (1937-38) 271-8.
   Through the hank
   Newspell
   Synthetic seconds
   Cigarette vs beer
   That quaint joker
   Red/black overhand shuffle
   Red/black riffle shuffle
   Color trouble
   Draw poker plus
   A smart cigarette
   An easy lesson
   Three piles
   The Houdini pen
   Pocket to cuff
   Fountain pen gone
   Card-minded

"Power." 41 (Feb 1938) 280. (A red-blue transposition)
"The ... of ... Trick." Summer Extra (1938) 323. (Card spell)
"Watchistry." 51 (Dec 1938) 357. (Divination effect)
"How to Make a Nazi Cross." 129 (Mar 1941) 746.
   (Match gag in Editrivia)
"Martin Gardner's Pen Kink." 132 (Apr 1941) 759.
   (Pen or pencil vanish)

Latter Day Secrets

"The Balancing Deck." 2 (2000) 25-6.
"Match Ace", match puzzle. 7 (2002) 264.
"Prediction." 8 (2003) 327-8.
"William Tell" and "Band Box." 10 (2005) 393-6.

The Linking Ring

Rising hat effect with photo of Gardner. **29**, 4 (Aug 1949) 34-5.
"The Incredible Dr. Jaks" (Article). **34**, 8 (Oct 1954) 23-5.
Handling for the Kraus Principle. **58**, 3 (Mar 1978) 87.
"Foursum." **58**, 4 (Apr 1978) 82-3.
   (Also a piece about Gardner, by John Braun, on pages 47-48.)
"Hocus Pocus Parade." **58**, 12 (Dec 1978) 67-85.
   (from Gardner's *Encyclopedia*)
"Five for Howard." **68**, 8 (Sept 1988) 89-91.

Looking Glass

"Mysterious Ring." 2 (Spring 1996) 59.

Magic

"Penny Tube." **2**, 6 (Feb 1993) 52.
"The Farmer and the Watermelon." **2**, 8 (Apr 1993).
"Band Release." **2**, 12 (Aug 1993).
"The Come-Back Bill." **3**, 2 (Oct 1993) 59-60.
"Pushdown." **3**, 5 (Jan 1994) 51-2.
"Gardner's Magnetized Knife." **3**, 8 (Apr 1994) 71.
"Triangular Curiosity." **3**, 11 (Jul 1994) 57-8.
"How to Make Magic Changing Pictures." **6**, 6 (Feb 1997) 60-4.
"Various Anecdota Magicana." **8**, 10 (Jun 1999) 112.
"Optimal Optical Cut." **8**, 12 (Aug 1999) 77-8.
"Newspaper Prediction." **15**, 12 (Aug 2005) 80.
"Shuffle." **16**, 9 (May 2006) 85.

"Can't Loose Change." **17**, 10 (Jun 2007) 86.
"Just Say Yes." **18**, 8 (Apr 2009) 5.
    Letter about Larry King and Criss Angel.
"Cups Ahoy." **19**, 6 (Feb 2010) 71-2.
Six physics tricks reprinted by Jason England. **19**, 11 (July 2010) 62-3.

Additional citations in the "Physics Tricks" chapter.

Magic Arts Journal

"Finger-Nose Do-As-I-Do." **2**, 2 (Mar 1988) 13.
"Restored Match." **2**, 4 (1988) 24-5. (A triple issue, nos. 4, 5, 6)

Magick

"Torn Corners." 172 (1977) 859-60.
"Keyword." 282 (1981) 1408. (with Phantini)

Chat in 137 (p.680) and 171 (p. 854).

Magic Magazine

"The Magic Prediction Square." **1**, 1 (Feb 1975) 20-1, 32.
    ("Martin Gardner's Puzzle Page"; a series over 3 issues).
"Mind Reading with Dice." **1**, 2 (Mar 1975) 17, 51.
"The Hummer Cup Trick." **1**, 3 (Apr 1975) 26, 34.

Magicol

"Editions of Erdnase." (Aug 1951) 4-5.
"My Chicago Booklets." 165 (Nov 2007) 16-7.
"Bob Hummer and 'Senator' Crandall." 168 (Aug 2008) 20-3.

The Magic Wand

"Sun and Moon for Children." **36**, 216 (Dec 1947) 164.
"Mathematics, Magic, and Mystery." **46**, 254 (1957) 50. (Excerpts.)

M–U–M

Handling of a trick of Bruce Elliot. **57**, 2 (Jul 1967) 20.
"Curious Speller." (Jun 2007) 66-7. (Written by Colm Mulcahy.)

The New Jinx

"Martin Gardner's Pen Kink." 26 (Jun 1964) 108.
    (Reprint from *Jinx* no. 132)

The Olram File

"Hide-out." 11 (1992) 4.

The New Pentagram

"Penetrating Paper Matches." **7**, 9 (Nov 1975) 69-70.
    (Short bio on p. 72.)

The New Phoenix

"A Silk Trick." 302 (5 Apr 1954) 7.
    (Trick by Dr. Daley; written by Gardner)
"Mity Match." 307 (14 May 1954) 25.
    ("Multiplying Spots", "Diminishing Spots")
"Airing the deck." 321 (24 Dec 1954) 92.
"Pencilvania." 334 (12 Dec 1955) 146.
"Knotty Problems." 368 (April 1962) 299-300.
    ("The Impossible Knot", "Knot Box")
"The Thirteenth Turn." 387 (Feb 1964) 375.
"Variable Magnetic Lie Detector." 393 (Nov 1964) 399-400.

Gardner is in the chat in numbers 316, 318, 321, 322, 329, 340, 391.

The Pallbearers Review

Version of the 5 by 5 problem. **3**, 4 (1968) 160.
"Odd and Even Dice." **3**, 9 (July 1968) 196.
    (A variation on *Phoenix*, no. 387.)
"Red and Black" (A biased wagering game). **4**,9 (July 1969) 272.
"Babel" column. **4**, 12 (Oct 1969) 286.
    (Note on Mutus, Dedit, Nomen effect.)

"Etcetera" column. **4**, 12 (Oct 1969) 288.
 (Handling tip for "Repeat Lie Speller.")
"Babel" column. **5**, 2 (Dec 1969) 304.
 (Anecdote about lost glove practical joke.)
"Etcetera" column. **5**, 4 (Feb 1970) 320.
 (Notes on Lyons "Numbers" trick.)
("Babel" column. **5**, 5 (Mar 1970) 324.
 (A biased card wagering game.)
"Mutus Nomen." **5**, 7 (May 1970) 338.
"Are You Psychic?." **5**, 9 (July 1970) 345.
"Turnabout Coins." **6**, 5 (Mar 1971) 412.
"Paradox Papers." **6**, 9 (July 1971) 429, 433.
"Color Thot." **7**, 4 (Feb 1972) 488.
"Three Rings." **7**, 7 (¿May 1972) 517.
"Missing Digit" from Gardner's Sam Loyd book. **8**, 2 (1972) 578.
"King James Version" (see also p. 780). **8**, 10 (Aug 1973) 643, 649.
A card vanish. **8** (Aug 1973) 665. (the Autumn "8th Folio.")
(From Gardner's review of John Fisher book). **9**, 2 (Dec 1973) 688.
"Martin Gardner's Parity Bet." **9**, 3 (Jan 1974) 700-1.
"Yin-Yang." **9**, 4 (Feb 1974) 710-1.
("Further Ideas"). **9**, 6 (Apr 1974) 726.
"Strange Bedfellows." **9**, 8 (1974) 749.
"Further Ideas" and slop shuffle variation. **9**, 9 (Jul 1974) 756.
"On 'Drop Out' ." **9**, 10 (Aug 1974) 782.
 (Also puzzle p. 776 and Coin to Cuff, p. 780
"With Magnets." **10**, 2 (Dec 1974) 829.
"He Does Tricks" presentation for Magic Tipper. **10**, 7 (May 1975) 962.
(Note on Kruskal Principle). **10**, 8 (June 1975) .
"Hypercard." **10**, 10 (Aug 1975) 1042.
"Penetrating Matches"). **10** (1976) 936. (Titled *Close-Up Folio No. 5.*

Pentagram

"The Shape of Things to Come." **13**, 11 (Aug 1949) 85.

Penumbra

"Stock Holdout." 9 (2005) 17. (credit correction in no. 11, p. 1)

The Phoenix

"Hand to Hand." 163 (5 Nov 1948) 655.

(Also a swindle from Senator Crandall)

"Die-Vination." 164 (19 Nov 1948) 658.

(Knuckle cracking gag). 164 (19 Nov 1948) 659.

"Ten Card Deal." 168 (1949) 672.

"Ten Card Deal P.S.." 170 (11 Feb 1949) 681.

"Beery." 188 (21 Oct 1949) 754. (Bar gag with salt and beer)

"Muddler." 196 (27 Jan 1950) 785. (Tie clip)

"It's a Natural." 199 (24 Mar 1950) 796.

(Card revelation with Bill Simon.)

(Practical joke involving milk). 203 (19 May 1950) 813.

"Anecdota Magicana." 216 (17 Nov 1950) 865.

(Stories deleted from *Argosy* article.)

"Spectacular." 234 (27 Jul 1951) 937.(In paragraph 12)

(A letter to Elliott reviewing Don Alan's act). 238 (21 Sep 1951) 953.

"Similar Twins." 241 (2 Nov 1951) 962.

(With Simon and a caricature by Dr. Jaks)

"McClenahan vs. the 'Astounding' Jim Walker." (2 Nov 1951) 964-5.

(About a fraudulent article the December 1950 issue of *Fate*)

"The Ruler." 244 (1951) 974.

"Easy Aces." 249 (22 Feb 1952) 997.

"Where Does the Bunny Go." 252 (4 Apr 1952) 1009.

(A novelty puzzle card from *Parents Magazine*, April 1952)

"Cubism." 254 (2 May 1952) 1016.

"Fir You." 271 (26 Dec 1952) 1082-3.

(Pyramidal paper Christmas tree)

"ESP and PK." 272 (9 Jan 1953) 1089. (From *In the Name of Science*, chapter 25)

Poker problem. 275 (20 Feb 1953) 1101.

(Answer in 6 March 1952, no. 276, p. 1104).

(Review of Hummer's "Three Pets"). 279 (17 Apr 1953) 1115.

(An invertible drawing). 279 (17 Apr 1953) 1116.

(Reprinted from *Humpty Dumpty*)

Gardner is in the chat in numbers 116, 160, 169, 190, 192, 196, 200, 201, 210, 212, 226, 227, 235, 236, 238, 243, 244, 245, 252, 272, 275, 276, 285, 293

Precursor

(A Si Stebbins card spell). 21 (Oct 1988).

Prolix

"A Tabletop Miser's Dream." 1 (2006) 38-9.
"Gardner's Gambit." 2 (2006) 69-79. (A parity game.)
"Band through Hand." 3 (2008) 186-7.
"Riddle rope" (with Schmidt). 4 (2008) 265.
"Trick Questions." 5 (2009) 297. (Mostly word-based quickies.)
"Berg-Hunter Knot." 6 (2009) 396.
"Two Deck Criss Cross." 9 (2011) 601.
"Jumping Skeleton Key." 9 (2011) 602.

Rigmarole

"Blow Up." 1 (1993) 10.
"Two Table Tricks ." 2 (1993) 13-4.(Vanishing Salt Shaker, Force Field)
"Hydrick's Bill Trick." 3 (1993) 30.
"Coiled." 4 (1993) 44.
"Crazy Coin." 9 (1994) 102.
"Rising marble." 10 (1994) 115.

S-C

"Dollar Bill Change." 1 (1985) 20.

The Sphinx

"New Color Divination." **29**, 3 (May 1930) 121.
"New Ring on String." **29**, 5 (Jul 1930) 214-5.
"The Best Pocket Tricks of Martin Gardner." **29**, 6 (Aug 1930) 240.
    (A pocket knife paddle move and a string and fountain pen effect)
"Fountain Pen Divination or Baffling Vision." **29**, 6 (Aug 1930) 255.
"An Ice Tea Effect." **29**, 7 (Sept 1930) 287.
"Call It What You Want Card Trick." **29**, 7 (Sept 1930) 287.
    (Also in this issue, a variation on Stebbins card coding).
"The Travelling Stick of Gum." **29**, 8 (Oct 1930) 319-20.
"Vanishing Pack of Life Savers." **29**, 8 (Oct 1930) 334-5.
"A New Egg Bag." **29**, 12 (Feb 1931) 519.
"An Impromptu Trick" (Coin transportation). **30**, 1 (Mar 1931) 20.
"Try These on Your Chewing Gum." **30**, 9 (Nov 1931) 406.
"Impromptu Divination." **32**, 3 (May 1933) 76.
"Parlor Mindreading with a Deck of Cards." **32**, 7 (Sept 1933) 200-1.

"Parlor Thought Transference." **32**, 7 (Sept 1933) 207, 211.
"Comedy Card Spelling Routine." **32**, 12 (Feb 1934) 371-2.
"Crayons and Hat." **32**, 12 (Feb 1934) 373-4.
"A Four Ace Routine that is Different." **33**, 1 (Mar 1934) 13-4.
"A Practical Coin Slight." **33**, 2 (Apr 1934) 45.
"Two Decks, Two People, and a Magician." **36**, 8 (Oct 1937) 218-9.
"Magic with Chessmen." **36**, 9 (¿Nov 1937) 251.
"Dice, Hat, and Matches." **39**, 2 (Apr 1940) 36-7.
     (He was the guest editor for this section.)
"Gardner's Goofy Silks." **39**, 10 (Dec 1940) 244.
"More Goofy Silks." **39**, 12 (Feb 1941) 322.
"Army Guts Test." **41**, 6 (Aug 1942) 118.
"Give Away Cards for Table Workers." **41**, 9 (Nov 1942) 178.
"The Magic of Dr. Jaks" (Article). **47**, 12 (Feb 1949) 334-7.
"The Vanishing Cigar." **51**, 3 (Dec 1952) 130.
     (Chat by Elliott on p. 132)

Swami [Calcutta]

"Dr. Jaks Predi-X-ion." **1**, 6 (1972) 22.
"Paper Fold Prediction." **2**, 19 (July 1973) 73-4.

The Swindle Sheet

"The Elusive Ace." 1 (1990) 6.
"Well-stacked." 2 (1990) 3. (with Fulves)

Underworld

"Cigar Levitation." 1 (1996) 1.
"Cornered." 2 (1996) 23.
"Face-Up Gemini Twins." 3 (1996) 29-30.
"Color Magnets." 4 (1997) 56.
"A Curious Number Force." 5 (1997) 62.
"Two Stunts." 7 (1998) 95.
"The K Diaries." 8 (1998) 103.
"The 7th Clip." 9 (1998) 120.
"Two Floating Cup Ideas." 10 (1999) 134-5.

Verbatim

"Fortune." 4 (1993) 25.

Xtra Credit

"Cherchez Again." 6 (2010) 30.
"2-way Flexagon." 8 (2010) 46.

PHOENIX

VOL. 18   FEBRUARY, 1937   No. 6

HENRY A. REESE
editor

WILBUR JERGER
business manager

AUDREY EICHENBAUM
art editor

EVERETT WARSHAWSKY......advertising mgr.
ELIZABETH McCASKEY..........circulation mgr.
WINSTON ASHLEY.................associate editor
ELIZABETH WESTON.............associate editor

EDITORIAL ASSOCIATES

V. P. Quinn   Theodora Schmitt
Margery Goodkind   William Crockett
C. Sharpless Hickman   Jean Garrigus
Dick Lindheim   Harvey Karlen
Meyer Becker   Martin Gardner
David Eisendrath   Sam Hair

BUSINESS ASSOCIATES

Harker Stanton   Eleanor Cupler
Harry M. Hess   Mary Ann Patrick
Dan Heindel   Betty Quinn
Buryl Lazar   Mary Letty Green
Robert Warfield   Bud Daniels
Mel Rosenfeld   Franklin Horwich

The University of Chicago official student magazine, established 1919.   Published monthly (except July and August) by and for the students
of the University of Chicago.   The University
is not responsible for any material herein, nor for
any contractual obligations.   Fifteen cents the
copy, one dollar the year from September to June.
Entered as second class matter, November 12, 1936,
at the post office in Chicago, Illinois, under the
Act of March 3, 1879.   The contents are not to
be printed without permission.   Mailing address,
Faculty Exchange, Box Ninety-seven University
of Chicago. Telephone Dorchester 7279.

CONTENTS

Last year Martin Gardner edited
Comment, gentle literary mother of
this year's Phoenix.   This year he
contributes to the erratic son.   He
seems to be a caricaturist, magician,
writer, and philosopher, though he
does everything so quietly that you can
never be sure—until he crashes
through with something like the Crane
Caricature on page 21.   Then you
know he's a caricaturist.   Sit around
until your watch disappears and you
know he's a magician.   Wait until he
gets a story in to Phoenix which he's
been promising for three months and
you'll know he's a writer.   Check up
on his registration and let him talk
privately to you of the Aristotelians,
and you know he's a philosopher.

He has an intense interest in his
surroundings and likes to wander
through old book-stores and North
Clark Street.   Hailing from Tulsa,
Oklahoma, he has somehow managed
to acquire a wide and varied acquaint-
anceship in Chicago.   If you want to
know of an obscure but excellent
place to eat, drink, or be otherwise
entertained, Gardner will know just
the place.

He's a small fellow with high nar-
row shoulders, an elastic mouth, and
deep-set, dark eyes.   That's he up
above.

About Martin Gardner

This appendix list all the known articles about Martin Gardner, which are predominantly interviews. If only a clipping was used, dates and page numbers may be missing. Further, previous attempts at indexing Gardner's work are cited in the last section.

A.1 Articles Based on Interviews

1. H.A.R., "Who is Gardner? What is He?" *Phoenix* [University of Chicago] **17** (June 1936) 15.

   The "University of Chicago's Monthly Humor Magazine"

2. No by-line, "[Martin Gardner]." *Phoenix* [University of Chicago] **18**, 6 (February 1937) 4.

3. No by-line, "Puzzles." *Tower Topics* [University of Chicago] (February 6, 1939) 2.

4. C. Sharpless Hickman, "Escape to Bohemia." *Pulse* [University of Chicago] **4**, 1 (October 1940) 16-17.

   The "University of Chicago's Student Magazine". There were two October 1940 issues.

5. No by-line, "Literally an Astonishing Young Man." *Tower Topics* [University of Chicago] (December 2, 1940) 1.

6. No by-line, *Esquire* (September 1949) 22.

A short biographical blurb with photo. See also October 1946, without photo.

7. LaVere Anderson, "Under the Reading Lamp." *Tulsa World–Sunday Magazine* (August 31, 1952) unknown (Section 5).
8. Don Morris, "The Shoebox Scholar." *University of Chicago Magazine* **45**, 5 (February 1953) 9-11.
9. LaVere Anderson, "Under the Reading Lamp." *Tulsa World–Sunday Magazine* (April 28, 1957) 28.
10. R. DeStefano, B. Appelman, S. Hartman, "An Interview with Martin Gardner." *Mathematics Student* [The Brooklyn Technical High School] **33**, 2 (June 1964) 4, 26.
11. Charlie Rice, "Romance by the Numbers." *The* [Baltimore] *Sun* (December 17, 1967) 267.

Calls up Martin Gardner and talks to Dr. Matrix

12. Bernard Sussman, "Exclusive Interview with Martin Gardner." *Southwind* [Miami-Dade Junior College] **3**, 1 (Fall 1968) 7-11.
13. [Martin Gardner], "Under the Reading Lamp." *Tulsa World–Sunday Magazine* (May 4, 1969) 18e.

Written by Gardner at the request of the columnist.

14. [Stefan Kanfer], "The Mathemagician." *Time* (April 21, 1975) 63.
15. Betsy Bliss, "Martin Gardner's Tongue-in-Cheek Science." *Chicago Daily News* (August 22, 1975) 27,29.
16. Brian McCallen, "His Fun Multiplies with Age." *Herald Statesman* [Hastings/Dobbs Ferry Section] (October 27, 1975) 3.
17. Philip Nobile, "Paranonsense." *Gastonia Gazette* (December 21, 1975) 5-B.
18. Hank Burchard, "The Puckish High Priest of Puzzles." *Washington Post* (March 11, 1976) 89.

   - [Hank Burchard], "Silver-Haired Scholar Spins a Skillful Web of Chance." *Winnipeg Free Press* (March 16, 1976) 13.

19. Scott Gordon, "Carrollians Rejoice! New 'Alice' Book Due." *Tulsa World* (March 21, 1969) B1-B2.
20. Sally Helgeson, "Every Day." *Bookletter* **3**, 8 (December 6, 1976) 3.
21. Irving Joshua Matrix [Peter Renz], "Martin Gardner: Defending the Honor of the Human Mind." *Two-Year College Mathematics Journal* **10** (1979) 227-231.

- Peter Renz, "Martin Gardner: Defending the Honor of the Human Mind." In *Mathematical Games* (Mathematical Association of America, 2005).

  Abridged and adapted version for use in the booklet accompanying the CD collection.

22. Anthony Barcellos, "A Conversation with Martin Gardner." *Two-Year College Mathematics Journal* **10** (1979) 232-244.

    - Anthony Barcellos, "Martin Gardner." In *Mathematical People*, edited by D. J. Albers and G. L. Alexanderson (Cambridge: Birkhauser, 1985) 99-107.
    - Anthony Barcellos, "Martin Gardner." In *Mathematical People: Second Edition*, edited by D. J. Albers and G. L. Alexanderson (A. K. Peters, 2008) 92-103.

23. Scot Morris, "Festschrift for the Master Gamesman." *Omni* **2**, 1 (1979) 176-177.

    - Scot Morris, "A Tribute to "The Mathemagician"." In *Omni Games*, edited by Scot Morris (Omni Publications, 1983) 68-69.
    - Scot Morris, "A Tribute to "The Mathemagician"." *Words Ways* (November 2014) 244-245.

24. Rudy Rucker, "Martin Gardner, Impresario of Mathematical Games." *Science81* **2**, 6 (July/ August 1981) 32-37.

    The title on the first page of the article is "Who Makes Math Marvelous, Turns Magic Satin Smooth, Tends the Looking-Glass Garden, and Can Make a Winner of Anyone Who Plays His Games?" with instructions for folding the page to leave only "Martin Gardner" showing.

    - Rudy Rucker, "Puzzle King Retiring to the Essay Life of a 'Scribbler'." *Chicago Tribune* (September 16. 1981). "Tempo" section

25. Jerry Adler and John Carey, "The Magician of Math." *Newsweek* (November 16, 1981) 101.

26. No by-line, "Calling All Martin Gardner Fans." *Focus* [American Mathematics Association] (November/December 1981).

27. Sara Lambert, "Martin Gardner: A Writer of Many Interests." *Time-News* [Hendersonville, NC] (December 5, 1981) 1,10.

28. Lynne Lucas, "The Math-e-magician of Hendersonville." *The Greenville* [South Carolina]*News* (December 9, 1981) 1B,2B.

29. Lee Dembart, "Magician of the Wonders of Numbers." *Los Angeles Times* (December 12, 1981) 1,10-21.

30. David Pope, "Meet the Mathemagician: Martin Gardner." *Games* (November/December 1981) 16-20.

    Contains large "Puzzler Sampler" inset, with 16 collected puzzles.

31. Scot Morris, "Interview: Martin Gardner." *Omni* **4**, 4 (January 1982) 66-69,80-86.

    • Scot Morris, "An Account of a Mathematician's Career." *Mathematics Teacher* **76**, 4 (April 1983) 276-282, 291.

32. Sara Lambert, "Martin Gardner – The Numbers Magician." *Time-News* [Hendersonville, NC] (February 24, 1982).

33. Sara Bingham, "Retirement Hasn't Stopped Gardner from Writing or Winning Awards." *Time-News* [Hendersonville, NC] (March 29, 1983) 1,12.

34. Art Levy, "Gardner Sees 'Threat to Science'." *Time-News* [Hendersonville, NC] (May 24, 1987) 1D.

35. Matthew J. Costello, "Martin Gardner—The Dean of American Puzzlers." In *The Greatest Puzzles of All Time* (Prentice Hall, 1988) 114-123. (Chapter 10) (also Dover, 2012)

36. Andy Duncan, "Master of Math Martin Gardner Relishes Role of Skeptic." *News and Record* [Greensboro, NC] (January 4, 1988) A8-A9.

37. Holly E. Selby, "The Magic Mathematician." *News and Observer* [Raleigh, NC] (March 6, 1988) 1C-2C.

38. T. H. Stickels//, "Having Fun with Math Games Has Been His Life's Work." *Irondequoit Press* [Rochester, NY] (November 16, 1992) 8-B.

    Was syndicated to other Messenger-Wolfe Newspapers.

39. Lawrence Toppman, "Mastermind." *The Charlotte* [NC] *Observer* (June 20, 1993) 1E, 6E.

40. Bob Fennell, "Gardner on Gardner: JPBM Communications Award Presentation." *Focus* [American Mathematics Association] (December 1994) 9, 33.

    Interview related to the award from the Joint Policy Board for Mathematics.

41. Pete Zamplas, "Science is Scintillating for 'Master of Puzzles'." *Time-News* [Hendersonville, NC] (May 13, 1996) 1A, 9A.

42. William Harmon, "The 1995 Tryon Festival and a Conversation with Martin Gardner." *North Carolina Literary Review* 5 (1996) 71-79.

43. Michael Shermer, "The Annotated Gardner." *Skeptic* **5**, 2 (1997) 56-61.

- *Martin Gardner 1914 - 2010, Founder of the Modern Skeptical Movement* (eSkeptic, 26 May 2010).

44. István Hargittai, "The Great Communicator of Mathematics and Other Games." *Mathematical Intelligencer* **19**, 4 (1997) 36-40.
45. Kendrick Frazier, "A Mind at Play." *Skeptical Inquirer* **22**, 2 (March/April 1998) 34-39.
46. Elizabeth Jennings, "A Visit with the Wizard." *Time-News* [Hendersonville, NC] (October 4, 1998) 1E, 5E.
47. Jan Susina, "Martin Gardner: The Annotator of Wonderland." *The Five Owls* **14**, 3 (January/February 2000) 62-64.
48. Richard Hatch, "Martin Gardner: In His Own Words." *Magic* (April 2000) 22-25.
49. Mark I. West, *Martin Gardner: North Carolina's Historian of Oz and Annotator of Alice.* North Carolina Literary Review
50. 10 (2001) 92-96.
51. Barbara Krasner-Khait, "Martin Gardner: Recreational Math Master." *Odyssey* **11**, 7 (October 2002) 22-24.
52. John Scott, "Annotating Alice." *Book and Magazine Collector* 246 (September 2004) 32-37.
53. Don Albers, "Master of Recreational Mathematics—and Much More: An Interview with Martin Gardner." *Focus: The Newsletter of the Mathematical Association of America* **24**, 8 (November 2004) 4-7.
54. Don Albers, "On the Way to 'Mathematical Games': Part I of an Interview with Martin Gardner." *College Mathematics Journal* **36**, 3 (May 2005) 178-190.
55. Don Albers, "'Mathematical Games' and Beyond: Part II of an Interview with Martin Gardner." *College Mathematics Journal* **36**, 4 (September 2005) 301-314.

Parts I and II were issued together as a booklet for the Gathering for Gardner 7.

- Don Albers, "An Interview with Martin Gardner." In *Impromptu*, edited by Todd Karr (Miracle Factory, 2015) 19-42. Parts I and II together.

56. James S. Tyree, "Gardner Cultivates Math Fun." *The Norman* [Oklahoma] *Transcript* (June 5, 2005) A1-2.
57. Allyn Jackson, "Interview with Martin Gardner." *Notices of the AMS* [American Mathematics Society] **52**, 6 (June/July, 2005) 602-611.

58. Richard Hatch, "Martin Gardner." *M–U–M* (January 2006) 77-78.
59. Peter Cannon, "The Magic of Martin Gardner." *Publisher Weekly* (September 25, 2006) 39-40.
60. Colm Mulcahy, "An Interview with Martin Gardner." *MAA Online* (November 6, 2006).

    www.maa.org/columns/colm/cardcol200610.html

61. Normand Baillargeon, "Martin Gardner : le sceptique polymathe." *Le Qubec sceptique* 59 (Spring 2006) 22-29.
62. Leigh Dayton, "The Face." *The Weekend Australian* [Review section] (February 3-4, 2007) 3.
63. Hailey Branson, "A Man of Math & Magic." *The Oklahoma Daily* [University of Oklahoma] (October 8, 2007) 1-2A.
64. Donald E. Simanek, "Mathemagician." *Make* **12** (2008) 80-86.

    In volume 13 Simanek continues with "Martin Gardner's Mathemagic," pp. 174-175, where he presents some methods (e.g. the Gilbreath principle) and tricks Gardner showed him.

65. No by-line, "Magicians Meet at Rivermont." *The Norman Transcript* (July 29, 2008) A8.
66. No by-line, "Interview with Martin Gardner." *The College Mathematics Journal* **40**, 3 (May 2009) 158-161.
67. John Peterson, "An Interview with Martin Gardner." *Gilbert Magazine* **12**, 7 (June 2009) 12-15.
68. John Tierney, "For Decades, Puzzling People with Mathematics." *New York Times* (October 20, 2009) D2.

    Also two entries on October 19 on tierneylab.blogs.nytimes.com

69. Hailey R. Branson, "Mathemagician." *Oklahoma Gazette* **31**, 49 (December 9, 2009) 21-23.
70. James D. Watts, Jr., "Oklahoma's Man of Letters—and Numbers." *Tulsa World* (April 4, 2010) G4.
71. Andrea Pitzer, *Martin Gardner, 1914-2010* (May 26, 2010) This tribute contains a 2008 interview and is found at hilobrow.com..
72. Joshua Jay, "Martin Gardner: An Interview." *Magic* **19**, 11 (July 2010) 58-61.
73. Kevin Thomas, "'I Grew Up on the Oz Books': An Interview with Martin Gardner." *Baum Bugle* **56**, 2 (Autumn 2012) 23-27.

A.2 Other Articles

1. LaVere Anderson, "Under the Reading Lamp." *Tulsa World–Sunday Magazine* (January 5, 1947) unknown (Section 5).
2. Lewis Nichols, "Puzzler." *New York Times Book Review* **66** (September 17, 1961) 8.
3. John Booth, "Memoirs of a Magician's Ghost." *The Linking Ring* **46**, 6 (June 1966) 37-39, 66-67.

   Chapter 13 of Booth's autobiography.

4. Peter Wilker, "[Martin Gardner]." *Hokus Pokus* **28**, 1 (1967) 4.

   A German magic periodical

5. Elaine Dewar, "In Search of the Mind's Eye." *Globe and Mail–Weekend Magazine* [Canadian] (July 30, 1977) 8-12.
6. John Braun, "Martin Gardner." *Linking Ring* **58**, 4 (April 1978) 47-48.

   • John Braun, "Martin Gardner." *Trick Talk* **4**, 4 (date unknown) [2-3].

7. No by-line, "Martin Gardner." *Contemporary Artists* **73-76** (1978) 226.
8. No by-line, "Martin Gardner." *Contemporary Authors* **24-26** (1978) 226.
9. No by-line, "Martin Gardner." In *The Encyclopedia of Science-Fiction*, edited by Peter Nichols (Dolphin, 1979) 244.
10. Anne Commire, "[Martin Gardner]." *Something About the Author* **16** (1979) 117-119.
11. Richard Buffum, "An Author Who Likes to Play Around with Magic." *Los Angeles Times* (July 2, 1981) OC-A3.

    Reads like a review of *Encyclopedia of Impromptu Magic* but it is not mentioned!

12. Doris Schattschneider, "Session in Honor of Martin Gardner." *American Mathematical Monthly* **89**, 5 (1982) C46.
13. Jonathan Austin, "Gardner Awarded AMS Prize." *Time-News* [Hendersonville, NC] (August 28, 1987).
14. Benjamin L. Schwartz, "Martin Garner: A Personal Reminiscence." *Journal of Recreational Mathematics* **22**, 1 (1990) 37-38.
15. Samuel Yates, "Impressions and Influences." *Journal of Recreational Mathematics* **22**, 1 (1990) 27-29.
16. No by-line, "Martin Gardner." *World Authors 1980–1985* (1991) 336-338.

17. Scott Power, "Mathematicians Honor Hendersonville Resident." *Time-News* [Hendersonville, NC] (August 14, 1992) 1, 15.
18. Douglas Hofstadter, *Martin Gardner: A Major Shaping Force in my Life* (1993). Talk given at the first Gathering for Gardner

    Published on the *Scientific American* website, May 24, 2010.

19. Richard Kostelanetz, "Martin Gardner." In *Dictionary of the Avant-Gardes* (a capella books, 1993) 84.
20. Prof M. O'Snart [Tom Ransom], "Martin Gardner – Encore!" *Magic* **3**, 8 (April 1994) 60-66.

    • Prof M. O'Snart [Tom Ransom], "Martin Gardner – Encore!" In *A Lifetime of Puzzles*, edited by Erik D. Demaine, et al (A. K. Peters, 2008) 15-27.

21. David Lister, "Martin Gardner and Paperfolding." *Fold* (1995).

    • David Lister, "Martin Gardner and Paperfolding." In *Mathematical Wizardry for a Gardner*, edited by Ed Pegg, Jr., Alan H. Schoen, and Tom Rodgers (A. K. Peters, 2009) 9-27.

22. Steve Jackson, "Grandmaster of Mathemagics." *Daily Telegraph* (August 19, 1995) 27 (Weekend Section).
23. Philip Yam, "The Mathematical Gamester." *Scientific American* (December 1995) 38-41.
24. Dana Richards, "Martin Gardner: A Documentary." In *The Mathemagician and Pied Puzzler*, edited by Elwyn Berlekamp and Tom Rodgers (A K Peters, 1999) 3-12. Some copies have covers labeled "Special Edition" but otherwise are the same.
25. M. C. [Matthew Creamer], "Martin Gardner." *Current Biography* **60**, 9 (September 1999) 30-33.
26. Richard Hatch, "Searching for Erdnase." *Magic* **9**, 4 (December 1999) 67-71.
27. No by-line, "The Ten Outstanding Skeptics of the Twentieth Century." *Skeptical Inquirer* **24**, 1 (January/February 2000) 24.
28. George P. Hansen, "Martin Gardner." In *The Trickster and the Paranormal* (Xlibris, 2001) 291-306.
29. Don Barry, "Gardner, Martin." In *Mathematics: Macmillan Science Library*, edited by Max Brandenberger (Macmillan Reference, 2002) 81-84.
30. Michael Shermer, "Hermits and Cranks." *Scientific American* (March 2002) 36-37.
31. Geoff Olson, "Martin Gardner: Genius Debunker." *Yukon News* (March 2004). now a blog page

32. [John Wilcock], "The Modest Genius of Martin Gardner." *The Ojai Orange* 36 (March 2005). ("John Wilcock's personal magazine")

   Mostly excerpts Kendrick Frazier's interview and a reprint of Gardner's "Fun with Möbius Band" from *Blackberries.*

33. Stefan Kanfer, *The Constant Gardner* (January 24, 2006). On his Gadflights blog

34. Luis A. Campistrous, Jorge M. Lopez, Celia R. Cabrera, "Martin Gardner y el Doctor Irving Joshua Matrix: Ficciones y Realidades." *Epsilon: Revista de la Sociedad Andaluza de Educacin Matemtica "Thales"* 67 (2007) 23-36.

35. Christopher Morgan, "Martin Gardner and His Influence on Magic." In *A Lifetime of Puzzles*, edited by Erik D. Demaine, et al (A. K. Peters, 2008) 3-13 .

36. Dana Richards, "The Gardner–Shaw Connection." *Friends of the Sherlock Holmes Collections* **13** (December 2009) 3, 8. (Published by the Special Collections Library at the Univ. of Minnesota.)

37. Greg Taylor, "How Martin Gardner Bamboozled the Skeptics." *Darklore* **5** (2010) 194-221.

38. Barry Cipra, "WWMD?" In *Raising Public Awareness of Mathematics*, edited by E. Behrends (Springer-Verlag, 2012) 331-337.

39. Michael Henle and Brian Hopkins, "Preface." In *Martin Gardner in the Twenty-First Century* (Math Assoc of America, 2012) v-vii.

40. Mark I. West, "Martin Gardner and His Contributions to the History of Oz." *Baum Bugle* **56**, 2 (Autumn 2012) 20-21.

41. Scott Cummings, "Martin Gardner's Oz Writings: A Checklist." *Baum Bugle* **56**, 2 (Autumn 2012) 22.

42. Antonio Peticov, *Antonio Peticov e o Universo de Martin Gardner* (EditoraLuste, 2013).

   97 page bilingual (Portuguese/English) book.

   • Antonio Peticov, *Antonio Peticov and the Universe of Martin Gardner* (Peticov.com, 2016).
   68 page book distributed at G4G12.

43. Jeremiah Farrell, Dana Richards, Thomas Rodgers, "Who was Armand T. Ringer?" *Word Ways* **49**, 2 (August 2016) 171-174.

   A review of Gardner's aliases.

44. Jeremiah Farrell, Karen Farrell, Thomas Rodgers, "Martin Gardner and Marilyn vos Savant." *Word Ways* **49**, 4 (November 2016) 244-251.

45. Chris Morgan, "Delaware and Dodgson, or, Wonderlands: Egad! Ado!" *Knight Letter* 99 (Fall 2017) 1-2.

    Detailed report on a talk about Gardner in the newsletter of the Lewis Carroll Society of North America.

46. Joshua Blu Buhs, *Martin Gardner as a(n Anti-) Fortean* (2017). Oblique Angle blog

A.3   Foreign Language Articles

1. "El Universo Ambidextro: Martin Gardner juega y entretiene con la antimateria y los laberintos." *La Opinion Cultural* [Argentinia] (June 7, 1975) 6-7.
2. No by-line, "Martin Gardner Anotado [Spanish]." *Snark* **2**, 7 (October/November 1977) 12-14.
3. "Garn geworfen." *Der Spiegel* (November 14, 1977) 254, 256-257.
4. Keiko Murata, "Magic Circle of Thought [Japanese]." *Objet Magazine Yu* (April 1979) 103-107.
5. Gian Marco Rinaldi, "I trucchi di Martin Gardner [Italian]." *Il Mondo della Parapsicologia* (February 1980) 69-71.
6. No by-line, "Martin Gardner: le mathemagicien [French]." *La Figaro Magazine* (February 23, 1980) 38.
7. [Izuo Sakane], "[Martin Gardner] [Japanese]." [Asahi newspaper] (1981).
8. Hans Holgen Anderson, "Profil: Martin Gardner [Danish]." *Skeptica: et kritisk forum for off-beat litteratur og pseudovidenskab* 2 (1981) 14-16.
9. Jaime Poniachik, Eduardo Abel Gimenez, "Martin Gardner o la segunda fundacion de los juegos matematicos." *Quid* [Argentina] **2**, 14 (1983) 162-164.

    • Jaime Poniachik, Eduardo Abel Gimenez, "El mago matemático de hoy [Spanish]." *Cacumen* **1**, 5 (June 1983) 7-9.

10. Marek Penszko, "Miedzy magia a filozofia [Polish]." *Problemy* 6 (1983) 56-57.
11. Marek Penszko, "Konkurs M. G. [Polish]." *Problemy* 7 (1983) 56-57.
12. Tom Werneck, "Freude am Knobeln und Denken [German]." *FR* [Frankfurter Ring] *Magazin* (December 17, 1983). Essentially a checklist of German translations.
13. J. van de Craats, "Van, Voor, en Over Martin Gardner [Dutch]." *Nieuw Archief voor Wiskunde: 4* **2**, 2 (1984) 311-316.

14. Stansilav Vejmola, "Martin Gardner a jeho matematické zábavy [Czechoslovak]." *Vesmír* (1988) 530.
15. Alain Zalmanski, "Martin Gardner [French]." *Jouer Jeux Mathématiques* 11 ((1991)) 4-5.
16. Warut Roonguthai, "Martin Gardner [Thai]." [Science Magazine] (July/August 1992) 251-252.
17. Bencze Gyula, "Meg kell védeni a tudás becsületét [Hungarian]." *Természet Világa* **125** (November 1994) 510-511.
18. Staar Gyula, "A matematika játékos lelke [Mathematics Player's Soul] [Hungarian]." *Természet Világa* **129** (February 1998) 64-67.

Contains interview.

19. Jesus de Paula Assis, "O anotador de Alice [Portugese]." *Folha de Sao Paulo* (June 14, 1998) 13.
20. Manuel Lopez Michelone, "Personaje notable [Spanish]." *PC Semanal* **15**, 355 (May 17, 1999) 16.
21. Trans. [Fara Di Maio], "Due parole con Martin Gardner a Proposito di Uriah Fuller [A few words with Martin Gardner about Uriah Fuller] [Italian]." In *Confessioni di un Medium*, edited by Massimo Polidoro (Padova: CICAP, 2006) 11-13. Number 1 in the series *I Quaderni di Magia*
22. Normand Baillargeon, "Martin Gardner: Le Sceptique Polymathe [French]." *Le Québec Sceptique* 59 (Printemps 2006) 22-29.
23. Ennio Peres, "Martin Gardner: Il Giocoliere della Matematica [The Juggler of Mathematics] [Italian]." In *Vite Matematiche*, edited by Claudio Bartocci et al (Springer / I Blu, 2007) 305-312.

A shortened translation appeared in 2011.

24. Alex van den Brandhof, *De Aha-Erlebnissen van Martin Gardner* [Dutch] (21 Oktober 2009). Kennislink.nl. (with a May 2010 update)
25. Ulf Persson, "Martin Gardner (1914-2010) [Swedish]." *Normat* **58**, 3 (2010) 1-3.
26. Bianucci Piero, "Martin Gardner (1914-2010) [Italian]." *La Stampa* (May 25, 2010) 41.
27. Staar Gyula, "[A Player Popularizing Science] [Hungarian]." *Természet Világa* **141**, 8 (August 2010).
28. Pedro Alegría [Ezquerra], Santiago Fernández, "Martin Gardner, el Mago de la Divulgación [Spanish] [The Wizard of Dissemination]." *La Gaceta de la Real Sociedad Matematica Española* **13**, 4 (2010) 671-704.

29. Pedro Alegría Ezquerra, "Magia y Matemáticas de la Mano de Martin [Spanish]." *Números* **76**, 4 (Marzo de 2011) 19-29.

30. Krzysztof Turzy'nski, "Gardner Kontra Pseudonauka: 60 Lat Walki [Polish]." *Delta* 1 (2011) 6-8.

A.4   Videos

1. *The Mystery and Magic of Mathematics: Martin Gardner and Friends* (Eugene, OR: New Dimensions Media, 1996).

   This 46 minute videotape was made from an hour-long episode of the series *The Nature of Things with David Suzuki*, titled "Martin Gardner: Mathemagician." Originally aired March 14, 1996 on the CBC Television Network, repeated January 2, 1997.

A.5   Obituaries and Tributes

1. No by-line, "Martin Gardner, 95, Math and Science Writer, Dies." *Associated Press* (May 23, 2010).

2. No by-line, "Martin Gardner." *Daily Telegraph* [UK] (May 25, 2010).

3. No by-line, "Martin Gardner." *The Economist* **395** (June 5, 2010) 94 .

4. No by-line, "In Memoriam." *The Mathematical Gazette* **94**, 530 (July 2010) 353. Short.

5. No by-line, "Martin Gardner; Journalist and Science Writer Who Made Mathematics Accessible to All and Decoded the Numerical Riddles in Lewis Carroll's Alice Books." *The* [London] *Times* (June 8, 2010) 57 .

6. No by-line, "Focus on . . . Martin Gardner." *Secondary Magazine* **62** (June 2010). unpaginated

7. No by-line, "Martin Gardner (1914-2010)." *Games* (October 2010) 2.

8. Don Albers, "Martin Gardner, 1914-2010: Magical Man of Numbers and Letters." *Math Horizons* **18**, 1 (September 2010) 5.

9. Normand Baillargeon, "Le Mathematician Martin Gardner n'est plus [French]." *Le Devoir* (26 Mai 2010).

10. Alex Bellos, "Martin Gardner Obituary." *Guardian* [Manchester] (May 27, 2010). See also letter on May 28.

11. Emma Brown, "Martin Gardner, 95: Inventive Author and 'Remarkable Eccentric Mind'." *Washington Post* (May 25, 2010) B5.

12. Mark Burstein (Ed), *A Bouquet for the Gardener: Martin Gardner Remembered* (The Lewis Carroll Society of North America, 2011). The contents are:

Foreword, Jim Gardner, ix-xi
Introduction, Mark Burstein xii-xiv
Martin Gardner & the Grown-Ups, Will Brooker, 3-5
My Pen Pal, Martin Gardner, Angelica Carpenter, 6-7
Martin Gardner: Ave Atque Vale, Morten N. Cohen, 8-11
Martin Gardner, an Appreciation, Selwyn Goodacre, 12-15
Perhaps a Few Wise Words: A Tribute to Martin Gardner, Edward Guiliano, 16-21
The Annotated Martin Gardner, Peter E. Hanff, 22-25
Dr.Matrix in Oz, Michael Patrick Hearn, 26-63
Martin Gardner, Major Shaping Force in My Life, Douglas Hofstadter, 64-73
The Stanford Gardner Archives, Stanley Isaacs, 74-77
Martin Gardner: Through the Looking Glass, Scott Kim, 78-82
Martin Gardner & Annotation, Jim Kincaid, 83-86
All in a Golden Afternoon, Charlie Lovett, 87-89
Gardner in Japan, Yoshiyuki Momma, 90-92
Martin Gardner & Lewis Carroll: The Magical Connection, Christopher Morgan, 93-96
Gardner in the U.K., Mark Richards, 97-98
Memories of Martin, David SChaeffer, 99
Martin Gardner: A Personal Reminiscence, Justin G. Schiller, 100-103
A Rememberance of a Gracious Carrollian, Byron Sewell, 104-106
Annotations of Immortality, Brian Sibley, 107-109
In the Garden of the Snark, Mahendra Singh, 110-113
Reminiscences, David Singmaster, 114-116
A Day in Norman, Alan Tannenbaum, 117-118
My Correspondence with Martin Gardner, Edward Wakeling, 119-125
Editing Martin, Robert Weil, 126-129
[two Gardner articles reprinted]
A Triad of Mathematical Popularizers, Francine F. Abeles, 150-159
One More Class: Martin Gardner & Logic Diagrams, A. Moktefi & A. Edwards, 160-174
Probability Paradoxes, Eugene Seneta, 175-183
Four Puzzles for Martin, Raymond Smullyan, 184-190
Charles Dodgson's Geometry, Robin Wilson, 191-204
A Carrollian Bibliography, August A. Imholtz, Jr., 207-231

13. Erik D. Demaine, "Recreational Computing." *American Scientist* **498**, 6 (November 1, 2010) 452-456.

14. Jeremiah Farrell, Karen Farrell, "80 Years of Gardner Magic." *Word Ways* **43**, 3 (August 2010) 105-106. Concentrates on magic and reprints "New Color Divination" (*Sphinx* 1930).

15. Chris French, "Martin Gardner: 1914–2010." *Guardian* [Manchester] (May 25, 2010).

16. [The Gardner Family], *Martin Gardner* (Norman Transcript, May 28, 2010).

17. Michael Gessel, "In Memoriam: Martin Gardner." *Baum Bugle* **54**, 2 (Autumn 2010) 40.

18. Ben Goldacre, "How Martin Gardner Warned Us to Beware the Bee People of Mars." *Guardian* [Manchester] (May 29, 2010). with a postscript by Wendy M Grossman

19. John W. Grafton, "Martin Gardner Remembered." *Dover Mathematics & Science* [Catalog] (Summer 2010) 32.

20. William Harmon, "Martin Gardner, 1914-2010." *Sewanee Review* **120**, 1 (Winter 2012) 134-137.

21. Michael Henle and Brian Hopkins (Eds), *Martin Gardner in the Twenty-First Century* (The Mathematical Association of America, 2012).

    The book collects eight articles written by Martin Gardner for *The College Mathematics Journal*, and various articles that were invited to be published in that journal to celebrate, comment on, or extend the ideas that Gardner had initiated over the years. Contributors are: Gillian Saenz, Christopher Jackson, Ryan Crumley, Greg W. Frederickson, Karl Schaffer, Richard J. Jerrard, John E. Wetzl, Arthur T. Benjamin, David Appplegate, Marc LeBrun, N. J. A. Sloane, Robert Bekes, Jean Pederson, Bin Sha, Sarah-Marie Belcastro, Tomasz Bartnicki, Jaroslaw, Grytczuk, H. A. Kierstead, Xuding Zhu, Ethan J. Berkove, Jeffrey P. Dumont, Ionut E. Iacob, T. Bruce McLean, Hua Wang, Les Pook, David Callan, Thomas Koshy, Z. Gao, Tiina Hohn, Andy Liu, Frederick V. Henle, John Beasley, Alexander Karabegov, Jason Holland, John J. Watkins, Ian Stewart, Brian Hopkins, Hsin-Po Wang, Owen O'Shea, Yossi Elran, Avierzi Fraenkel, Christopher N. Swanson, Stephen Lucas, Jason Rosenhouse, Andrew Schepler, Darren Glass, Jorge Moraleda, David G. Stork, Tanya Khovanova, John Stillwell, Clement Falbo, and Philip Straffin.

22. Harold R. Jacobs, "A Tribute to Martin Gardner." *Word Ways* **43**, 3 (August 2010) 170-171.

23. Douglas Martin, "Martin Gardner, Puzzler and Polymath, Dies at 95." *New York Times* (May 23 2010).

24. Thomas Maugh II, "Martin Gardner Dies at 95: Prolific Mathematics Columnist for Scientific American." *Los Angeles Times* (May 26, 2010).

25. Max Maven, "In Memorium: Martin Gardner 1914-2010." *Genii* **73** (July 2010) 24-25.

26. Marvin Miller, "The Martin Gardner I Knew." *Magic* **19**, 11 (56-58).

27. Stephen Mirsky, *Scholars and Others Pay Tribute to "Mathematical Games" Columnist Martin Gardner* (May 24, 2010). *Scientific American* website.

    John Allen Paulos, Douglas Hofstadter, Dennis Shasha, and Ian Stewart quoted

28. Ennio Peres, "Martin Gardner: The Mathematical Jester." In *Mathematical Lives*, edited by Claudio Bartocci et al (Springer, 2011) 217-220. See also Foreign Language Articles.

29. Jon Racherbaumer, "On the Slant." *Genii* (August 2010).

    • Jon Racherbaumer, "By the Numbers." *Word Ways* **43**, 3 (August 2010) 163-164.

30. Dana Richards, "Martin Gardner (1914–2010)." *Science* **465** (June 17, 2010) 884.

31. Rudy Rucker, "Martin Gardner." *Time* (June 7, 2010) 21.

32. Morton Schatzman, "Martin Gardner: Scientific and Philosophical Writer Celebrated for His Ingenious Mathematical Puzzles and Games." *The Independent* (May 29, 2010).

33. Howard Schneider, "Martin Gardner: The Polymath." *The Humanist* **329** (July/August 2010) 46-47.

34. Marjorie Senechal, "Martin Gardner (1914–2010)." *Mathematical Intelligencer* **33** (March 2011) 51-54.

    With contribution from: Alice and Klaus Peters, Ian Stewart, Alexander Shen, Ravi Vakal, Robin Wilson, and David Rowe.

35. Jorge-Nuno Silva, "Martin Gardner (1914–2010)." *EMS* [European Mathematical Society] *Newletter* 79 (March, 2011) 21-23.

36. David Singmaster, "Martin Gardner (1914–2010)." *Nature* **329** (July 9, 2010) 157.

37. [Various authors], "Martin Gardner." *Skeptical Inquirer* **34**, 5 (September/October 2010) 28-42.

    Ray Hyman, "Martin Gardner: A Polymath to the Nth Power"
    James Randi, "Martin Gardner Has Left Us"
    Paul Kurtz, "Martin Gardner's Contributions to the World of Books"

James Alcock, "We Have Lost an Icon"
Kendrick Frazier, "A World Treasure"
Joe Nickell, "Martin Gardner's Presence"
Robert Scaeffer, "The Humble Demigod"
Robert Carroll, "Exposing Crackpots and Charlatans"
Bryan Farha, "Visits to Martin"
John Allen Paulos, "The Connoisseur of Paradox"
Scott O. Lilienfeld, "Characterizing the Hermit Scientist"
Christopher C. French, "The Friend I never Met"
Neil deGrasse Tyson, "The Last of the Polymaths"
Jay M. Pasachoff, "The Roots of Skepticism"
Martin Bridgstock, "A Blowtorch tTurned on Jell-o"
Luis Alfonso Gamez, "Good-bye, Master Journalist"
Benjamin Radford, "What Martin Taught Me"
Timothy Binga, "My Reminiscence of Martin Gardner"

38. [Various Authors], "Remembering Martin Gardner." *MAA Focus*
    30, 4 (August/September 2010) 12-13.

By Peter Renz, John Derbyshire, Don Albers and Ian Stewart

There have been many (too many to list them all) tribute blog
entries. Here are some of the more substantial ones from 2010:

Anthony Barcello	The Back Bench	May 23
J. L. Bell	Oz and Ends	May 24
Tom Braunlich	united states chess federation	May 28
William Briggs	wmbriggs	May 25
Loren Coleman	Cryptomundo	May 23
Andrew Cusack	own blog	June 6
Richard Dawkins	richarddawkins	May 23
"Dean"	The Chicago Blog	May 24
John Derbyshire	the corner—National Review	May 24
David Elzey	the excelsior file	May 24
Lance Fortnow	Computational Complexity	May 24
Kendrick Frazier	CSICOP	May 24
Brian Hayes	bit-player	May 24
Chris Hallquist	uncredible hallq	May 29
Michael P. Hearn	Matilda Joslyn Gage Foundation	May 31
Jeff Hecht	Culture Lab—NewScientist	May 24
Konrad Jacobs	Wired	June 2
Stefan Kanfer	city-journal	May 27
Daniel Loxton	skepticblog	May 25
Roger Kimball	Roger's Rule	May 23

Thaiany Mota	Virtual Thaiany	May 25
Colm Mulcahy	Spelman.edu/c̃olm	June 14
Bill Mullins	NexusForum	May 24
Pradeep Mutalik	Wordplay–NY Times	May 31
Steven Novella	skepticblog	May 24
Tom Noyes	TommyWonk	May 24
Matt Parker	Eureka Zone—The Times	May 24
Ed Pegg, Jr	WolframAlpha Blog	May 27
Ivars Peterson	Mathematical Tourist	May 23
James Randi	JREF	May 22
Rudy Rucker	Rudy's Blog	May 24
Michael Shermer	True/Slant	May 24
Terry Stickels	own blog	June 7
Jan Susina	Ghost of the Talking Cricket	May 31
Tanya Thompson	thinkfun.com/puzzlehunter	May 31
Bobby Warren	IBM, magician.org	July 7
Colin Wright	Colins Blog	May 22
Kim Yu	Xiamen network [in Chinese]	May 30
No by-line	LCSNA	May 23
No by-line	maa.org/news	May 24

A.6 Centenary Articles

The year 2014 started with the announcement of the Gardner theme for Mathematics Awareness Month ("Mathematics, Magic, and Mystery," April 2014). In addition to all the articles a video was put on-line of Martin Gardner receiving the MAA Communications Award in 1994 (14-minutes, released 15 Oct 2014).

1. Tereza Bartlova, "Martin Gardner – Ke Stemu Vyroci Narozeni [Czech]." *Pokroky Matematiky,Fysiky a Astronomie* **59**, 2 (2014) 146-160.

   - Tereza Bartlova, "Martin Gardner's Mathematical Life." *Recreational Mathematics Magazine* 2 (September 2014) 21-40.

2. Elwyn Berlekamp, "The Mathematical Legacy of Martin Gardner." *SIAM News* **47**, 7 (September 2014) 4,7.

3. Fernando Blasco, "El universo matemgico de Martin Gardner: Juegos, acertijos, paradojas y otras maravillas recreativas [Spanish]." *Temas* 77 (Julio/Septiembre 2014).

4. James Case, "Martin Gardners Mathematical Grapevine." *SIAM News* **47**, 3 (April 2014).

5. Bob Crease, "Celebrating the Mind." *Physics World* (October 2014) 20.

6. Tom Ewing, "Martin Gardner: A Celebration of his (12 + 3 - 4 + 5 + 67 + 8 + 9)th Year." *M-U-M* (October 2014) 3640.

   Featuring interviews with Mark Setteducati, Dana Richards, Max Maven, John Railing, James Randi, Dick Hatch.

7. Jeremiah Farrell, "A Tribute to 'Armand T. Ringer'." *Word Ways* **47**, 4 (November 2014) 242-243. A crossword.

8. John Gillaspy, *Martin Gardner: 100 Years of the Magic of Physics* **52**, 7 (October 2014).

9. Jim Henle, *Celebrity Chefs* **36**, 3 (2014).

   Finds commonality between Julia Child and Martin Gardner through public involvement. See also his *The Proof and the Pudding* (Princeton University Press, 2015).

10. Max Maven, "A Gardnerian Centennial: A Brief Look Back at Martin Gardner." *Genii* (October 2014) 31.

11. Colm Mulcahy, "The Top 10 Things Every Mathematics Student Should Know about Martin Gardner." *Math Horizons* **22**, 3 (September 2014) 24-25.

    Also a review of *Undiluted Hocus-Pocus*.

12. Colm Mulcahy, "In Praise of a Deservedly Popular Mathematician." *Atlantic Journal and Constitution* (October 21, 2014).

13. Colm Mulcahy, "A Man of Number and Letters." *Norman* [OK] *Transcript* (October 25, 2014) A4.

14. Colm Mulcahy, Al Goetz, "The Best Friend Mathematics Ever Had." *Mathematics Teacher* **108**, 3 (October 2014) 194199.

15. Colm Mulcahy, Dana Richards, "Let the Games Continue." *Scientific American* **311**, 4 (October 2014) 9095.

    • Colm Mulcahy, Dana Richards, "Cien aos con Martin Gardner." *Investigacin y Ciencia* 457 (October 2014).

      Also a Polish, Japanese and two Chinese editions

    • Colm Mulcahy, Dana Richards, "Let the Games Continue." In *The Best Writing on Mathematics: 2015*, edited by Mircea Pitici (Princeton University Press, 2016) 14-25.

16. "Pictures from the Past." *Linking Ring* **94**, 10 (October 1, 2014).

17. Ivars Peterson, "Honoring a Century of Martin Gardner." *MAA FOCUS* (October/November 2014) 1213.

18. Dana Richards, "Martin Gardner: Further Explorations in Oz." *Baum Bugle* **58**, 2 (Autumn 2014).

19. Adam D. Rigny, . Notices of the AMS
20. *Mathematician, Magician, Mysterian* (April 2014).
21. Raymond Simon, "In Memory of Martin Gardner." *Games World of Puzzles* (December 2014) 48.
22. "El Universo Matemágico de Martin Gardner." *Temas 77* (Third Trimestre 2014).

This a glossy special issue of a periodical issued by *Investigacion y Cienca* (the Spanish edition of *Scientific American*) devoted to Martin Gardner. It reprints 14 columns, with a new introduction by Fernando Blasco.

Skeptical Inquirer ("In Celebration of Martin Gardner" and free access to classic "Notes of a Psi-Watcher" columns: "Lessons of a Landmark PK Hoax," "Facilitated Communication and False Memory," and others, in November/December 2014 issue, 15 October 2014.

The *Scientific American* website, on September 16, 2014, celebrated the Centennial with various pages posted as follow-ons to the August article:

- "Make Your Own Hexaflexagon ... and Snap Pictures of Them"
- "The Number 2,187 Is Lucky. Here's Why"
- "Can You Solve a Puzzle Unsolved Since 1996?"
- "How Well Do You Know Martin Gardner? [Quiz]"
- "Test Yourself: Try Some Classic Recreational Math Puzzles"

Various Centennial blogs and postings are listed below.

Aziz Anan	Celebration of Mind	10/21/2014
Gary Antonick	NYT Wordplay/Numberplay	10/13/2014
Gary Antonick	NYT Wordplay/Numberplay	10/20/2014
Normand Baillargeon	Chroniques/prise-de-tete	3/18/2015
Alex Bellos	Guardian blog (answers 10/27)	10/21/2014
Tim Binga	Center for Inquiry:Access Points	10/14/014
Breen and Emerson	AMS's Math Digest	10/1/2014
Fernando Blasco	Investigacion y Ciencia [Spanish]	10/3/2014
Mark Carson	FatBrain	10/21/2014
Dilip D'Souza	Livemint	10/24/2014
Jason English	Cosmos Magazine	2/24/2014
Jason English	Magic Magazine at M360	10/21/2014

Jesper Mads Eriksen	MitFyn [Danish]	10/15/2014
Fortnow and Gasarch	Computational Complexity	10/21/2014
Frederic Friedel	Chess News blog	10/21/2014
Evelyn Lamb	Scientific American Observation	10/19/2014
Colm Mulcahy	Scientific American Guest Blog	4/16/2014
Colm Mulcahy	plus.maths.org	10/20/2014
Colm Mulcahy	BBC News Magazine	10/21/2014
	also 10/23/2014 and in Spanish on 10/26	
Colm Mulcahy	Huffington Post	10/28/2014
Colm Mulcahy	Card Colm	10/31/2014
Joseph Malkevitch	AMS Feature Column	4/1 2014
Burkard Polster	The Laborastory	9/28/2014
Polster and Ross	qedcat: The Age	10/20/2014
David Orden	Cifras y Teclas	10/21/2014
James Randi	JREF: James Randi Ed. Found.	10/14/2014
Will Shortz	NPR: Weekend Puzzler	10/19/2014
Simon and Devlin	NPR: Weekend Edition	4/12/2014
Greg Taylor	Daily Grail	10/22/2014

A.7 'Gathering for Gardner' Articles

The Gathering for Gardner (G4G) is a biennial event for about 400 people that is invitation-only. It was usually run by Tom Rodgers. People come to tell each other what they would like to tell Martin Gardner.

1. Alex Bellos, "The Science of Fun." *The Guardian, Weekend Section* (May 31, 2008) 32-39.
2. Alex Bellos, "Mathemagical." *New Scientist* (May 29, 2010) 42-47.
3. Arthur Benjamin, "Gathering for Gardner II." *Skeptic* **4**, 2 (1995) 16.
4. Robert P. Crease, "A Gathering for Gardner." *PhysicsWorld* (July 1, 2008). on-line only.
5. Robert P. Crease, "Gathering for Gardner: Homage to the Iconic Author of Scientific American's 'Mathematical Games' Column." *Wall Street Journal* (April 2, 2010) W11.
6. Bryan Farha, "Remembering Martin Gardner." *Skeptical Inquirer* **43**, 3 (May/Jun, 2019) 14.

7. Bo Emerson, "Fans of Math Icon Martin Gardner Gather in Atlanta." *Atlanta Journal and Constitution* (March 18, 2014).

8. Dan Garrett, "Pandora's Box: Martin-ized." *M.U.M.* **83** (June 1993) 613.

9. Dan Garrett, "G4G2: The Gathering for Gardner II." *Genii* **59**, 5 (March 1996) 346-348. also posted on Electronic Grymoire #782

10. Dan Garrett, "G4G3: January Convention Report." *Genii* **61**, 3 (January 1998) 52-53. also posted on Electronic Grymoire #1240

11. Ray Hyman, "The Eighth Gathering for (Martin) Gardner." *Skeptical Inquirer* **32**, 5 (September/October 2008) 12-14. with an afterword by Martin Gardner (p. 15)

12. Infinite Jest, "Encountering the Threshold of Humankind's Capacity for Mathematical Gamesmanship, at the Annual Gathering for Gardner." *Seed* (June 16 2006). on-line only.

13. Katharine Merow, "Gathering for Logic Puzzles." *MAAA Focus* (December/January 2011/2012) 4.

14. Scot Morris, "A Gathering for Gardner." *Omni* (June 1993) 96.

15. Joe Nickell, "Gardnerfest: Admirers 'Gather for Gardner' to Fete the Modest Genius." *Skeptical Inquirer* **20**, 3 (May/June 1996) 5-6.

16. Gábor Szabo Péter, *G4G7 – Atlanta 2006 [Hungarian]* **137**, 6 (Termeszet Vilaga Magazin, June, 2006).

17. Dana Richards, *Advances: Meeting of the Puzzlers* (Scientific American, June, 2014).

    Also on the webpage with the byline "George Groth" and the June *Investigación y Ciencia*.

18. Siobahn Roberts, "The Curious Side of Big Math." *Toronto Star* (April 30, 2006) D4.

19. Edward Rothstein, "Puzzles + Math = Magic." *New York Times* (April 3, 2004) B9, B11.

20. Edward Rothstein, "Puzzles, Origami and Other Mind-Twisters." *New York Times* (April 3, 2006).

21. G. Shapiro, "Puzzlers' Parley in Atlanta." *New York Sun* (March 25, 2004) 24.
22. G. Shapiro, "Tricksters Take Atlanta in Honor of Gardner." *New York Sun* (April 2, 2004) 14.

The focus is on Jay Marshall

23. Adam Rubin, "G4G9." *Magic* (May 2010) 29.
24. Bruce Torrence, "Gathering for Gardner." *Math Horizons* **18**, 1 (September 2010) 9-12.

There are also numerous unedited webpages and/or blogs that mention the Gatherings. Here are a few that are more focused on the Gatherings:

1. Robert P. Crease, *Martin Gardner Would Have Smiled* (April 16, 2018).

   physicsworld.com/a/martin-gardner-would-have-smiled/

2. Ivars Peterson, *A Gathering for Gardner* (March 27, 2006).

   www.maa.org/mathland/mathtrek_03_27_06

3. Ed Pegg Jr., *G4G7* (April 10, 2006).

   www.maa.org/editorial/mathgames/mathgames_04_10_06

4. Neil J. A. Sloane, *Notes on G4G7: Gathering for Gardner 7* (2006).

   www2.research.att.com/ njas/doc/G4G7

5. Erik Hermansen, *My First DROD Speech* (2006).

   forum.caravelgames.com

6. Katharine Merow, *Martin Gardner: In His Own Words* (October 16, 2014). maa.org

7. Jason Rosenhouse, *Gathering for Gardner 9* (April 1, 2010).

   scienceblogs.com/evolutionblog/2010/04/gathering_for_gardner_9

8. Tanya Thompson, *Good Times at G4G9* (May 20, 2010). thinkfun.com/puzzlehunter

   Also "More G4G9"

The conference has a tradition of gift exchanges, which was already a tradition at related meetings (like the International Puzzle Party). Attendees who participate must bring enough for all those participating (usually hundreds). Some are artifacts (like physical puzzles) and others are articles or contributions that can be printed on paper. The first event (1993), even before the gift exchange was formalized, distributed a cerlox-bound photocopy of such things but after that these were unbound. Starting with the seventh conference these were issued as attractive two-volume books. It is tempting to call these "proceedings" but they are not necessarily linked to presentations. The seventh, eighth, tenth, eleventh, twelfth, thirteenth and fourteenth have appeared and are shared only with those participating in the exchange; the ninth conference does not yet have one. Also see Festschrifts.

A.8 Festschrifts

In addition the Centennial volumes listed above there are several festschrifts that have appeared. Many of them are drawn from the various G4G presentations. Occasionally the papers are about Gardner.

1. Klarner, David A., Ed, *The Mathematical Gardner* (Wadsworth International, 1981).

   - Klarner, David A., *Mathematical Recreations: A Collection in Honor of Martin Gardner* (Dover, 1998).

2. Berlekamp, Elwyn, Rodgers, Tom, Eds., *The Mathematician and the Pied Puzzler* (A. K. Peters, 1999).

   Based on the First G4G. A separate "Special Edition" also appeared, that seems the same.

3. Wolfe, David, Rodgers, Tom, Eds., *Puzzlers' Tribute: A Feast for the Mind* (A. K. Peters, 2002).

   Based on G4G2, G4G3, G4G4.

4. Cipra, Barry, Demaine, Erik D., Demaine, Martin L., Rodgers, Tom, Eds., *Tribute to a Mathematician* (A. K. Peters, 2005).

   Based on G4G5.

5. Demaine, Erik D., Demaine, Martin L., Rodgers, Tom, Eds., *A Lifetime of Puzzles* (A. K. Peters, 2008).

   Based on G4G6. For Gardner's 90th birthday.

6. Pegg Jr., Ed, Schoen, Alan H., Rodgers, Tom, , Eds., *Homage to a Pied Puzzler* (A. K. Peters, 2009).

   Based on G4G7.

7. Pegg Jr., Schoen, Alan H., Rodgers, Tom, Eds., *Mathematical Wizardry for a Gardner* (A. K. Peters, 2009).

   Also based on G4G7.

8. Plambeck, Thane, Rokicki, Tomas, *Barrycades and Septoku* (AMS / MAA, 2020). Volume 100 in Spectrum series. Copyright held by G4G Inc. Based on later G4Gs.

A.9   Previous *Scientific American* Indexes

These citations are for previous attempts at indexing the column. Their number indicates that many people feel the need for such indexes. None compare to the indexing found here.

1. Dunn, James A., "A Martin Bibliography." *MT: Mathematics Teaching* 52 (Autumn 1970) 59-60.

   A list of the titles of the columns up to December 1969.

2. Dunn, James A., "A Martin Bibliography." *MT: Mathematics Teaching* 70 (March 1975) 58-59.

   A list of the titles of the columns from January 1970 to December 1974. A discussion (by D[avid] S. F[ielker]) of Gardner's related book is appended.

3. Schaaf, William Leonard, "Chronological Synopsis of Martin Gardner's Column in Scientific American." In *A Bibliography of Recreational Mathematics: Vol. 4* (National Council of Teachers of Mathematics, 1978) 138-144.

   A list of the columns up to August 1977, with very short descriptions. Based on Dunn's list.

4. Editors of Scientific American, "Index to Mathematical Games." In *Scientific American: Cumulative Index: 1948-1978* (Scientific American Inc., 1979) 397-400.

   A three-column topics index, to the individual issues up to Jun 1978.

5. Upton, Leslie J., *Index—Martin Gardner's Books* (1985).

   An 8 page photocopied cross-index of columns and chapters, as well as a supplement to Schaaf's list.

6. Lieske, Spencer, *Index of Martin Gardner's Books of Mathematical Games* (1980).

   A 5 page topics index, of the book chapters, used as a handout when speaking to math teachers

7. Lee, Carl W., *Gardner Index* (1987).

   This is a topics index, of the book chapters, that was made available electronically. It includes only the first 11 books.

8. Editors of Scientific American, "Index to Mathematical Games, Metamagical Themas and Computer Recreations." In *Scientific American: Cumulative Index: 1978-1988* (Scientific American Inc., 1989) 213-215.

   A continuation of the 1979 index, including the successor columns.

9. Singmaster, David, *Index to Martin Gardner's Columns and Books* (Technical Report SBU-CISM-95-09, South Bank Univ, London, 1995).

   This is a revision of his 1993 manuscript "Index to Martin Gardner's Columns and Cross Reference to his Books." It is updates the Schaaf and *Scientific American* lists above.

10. Liu, Andy, "Index of Martin Gardner's Scientific American Columns Anthologies." *Delta-K* ([MCATA—Mathematics Council of the Alberta Teachers' Association], 1996). Special issue "Mathematics for Gifted Students II."

    This contains an index of columns to chapters as well as a list of the separate problems.

11. Lee, Carl W., Charles Kluepfel, *Gardner Index* (1997).

    This 25 page topics index updates Lee's list, to include the final 4 books. It can be found at:
    www.ms.uky.edu/ lee/ma502/gardner5/gardner5.html

12. Miller, John, *Index to Mathematical Games* (1999).

    This column listing was started in 1975 but only distributed on the internet in 1999. It has been enhanced and can now be found at: //www.martin-gardner.org/MGSAindex.html

13. Langford, David, *Martin Gardner: Mathematical Game Collection* (n.d.).

   This webpage lists the titles to all the chapters. It can be found at www.ansible.uk /misc/mgardner.html

14. Simpson, David G., *Index to Martin Gardner's* Mathematical Games (n.d.).

   This webpage lists the titles to all the chapters, in original publication order; davidgsimpson.com.

15. Wikipedia, *List of Martin Gardner Mathematical Game Columns* (n.d.).

   This webpage lists titles of every *Scientific American* column. It can be found at:en.wikipedia.org/wiki/ List_of_Martin_Gardner_Mathematical_Games_columns

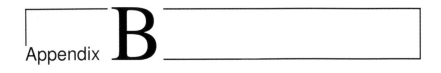

Appendix **B**

Foreign Language Editions

Introduction

Martin Gardner has been well-known for his varied and prolific writings. However it is usually not appreciated that his influence stretches around the globe. For over 60 years foreign editions of his books have been appearing and now they can be read in over 30 languages.

This bibliography contains about 400 translations, not counting variant editions. Most of these are separate efforts by translators. Occasionally there are even alternate translations of the same book into the same language. The various collections of the "Mathematical Games" columns from *Scientific American* account for about 70 of the translations, which is to be expected. What is surprising is that the influential *Fads and Fallacies* has only been translated five times. On the other hand, his science books *Relativity for the Millions* and *The Ambidextrous Universe* have been translated at least 16 times each. The daunting task of translating *The Annotated Alice* has been attempted twelve times.

Languages represented in this bibliography are:

Basque, British, Bulgarian, Catalan, Chinese, Czech, Danish, Dutch, Estonian, Finnish, French, German, Greek, Hebrew, Icelandic, Italian, Japanese, Korean, Latvian, Malayalam, Norwegian, Persian, Polish, Portuguese, Romanian, Russian, Slovak, Spanish, Swedish, Teluga, Turkish, Vietnamese

This list represents all the foreign editions that are known, but there are certainly many others that have escaped our attention. (The British editions are included in this bibliography; this reduces the redundancy in the main bibliography which only includes British editions if they are distinct or have priority.) The categorization and ordering of the books within each language is the same as in the main bibliography. We say "from" when referring to the book being translated because often parts were omitted, for whatever reason. This appendix only addresses books; articles or story translations are in the main bibliography above.

The asterisk (*) after a date is used to designate those editions that were only verified indirectly (for example found on OCLC). Multiple dates may correspond to different covers only; copies were not at hand to be inspected (just about half of these are in my collection).

B.1   Basque [Euskera]

Science

1. Trans. [unknown], *Zientzia eta Kultura: Idazki Zientifiko Hobere-nen Bilduma* [Science and Culture: A Collection of the Best Scientific Writings] (San Sebastián: Gaiak Argitaldaria, 1996*). From *Great Essays in Science*

B.2   British

Recreational Mathematics

1. *Mathematical Puzzles and Diversions* (George Bell & Sons, 1961*).
   - *Mathematical Puzzles and Diversions* (Penguin/Pelican, 1966, 1970*, 1973*, 1988, 1990, 1991*).

2. *More Mathematical Puzzles and Diversions* (Bell, 1963). From *The Second Book of Mathematical Puzzles and Diversions*
   - *More Mathematical Puzzles and Diversions* (Penguin/Pelican, 1966, 1969*, 1977*, 1990*).

3. *New Mathematical Diversions* (George Allen & Unwin, 1969*).
4. *Further Mathematical Diversions* (George Allen & Unwin, 1970*, 1975*). From *The Unexpected Hanging*

   • *Further Mathematical Diversions* (Penguin, 1977).

5. *Mathematical Carnival* (George Allen & Unwin, 1976*).

   • *Mathematical Carnival* (Penguin, 1978, 1990).

6. *Mathematical Magic Show* (George Allen & Unwin, 1978*).

   • *Mathematical Magic Show* (Viking, 1984*).
   • *Mathematical Magic Show* (Penguin, 1985).

7. *Mathematical Circus* (Allen Lane, 1981).

   Also published by Allen Lane for Book Club Associates

   • *Mathematical Circus* (Penguin, 1981, 1990).

8. *Science Fiction Puzzle Tales* (Penguin, 1983).
9. *Puzzles from Other Worlds* (Oxford University Press, 1986*).
10. Silvanus P. Thompson and Martin Gardner, *Calculus Made Easy* (Palgrave Macmillan, 1999*).
11. Boris Kordemsky, *The Moscow Puzzles: 539 Mathematical Recreations* (Penguin, 1975, 1990*). Edited by Gardner
12. H.E. Dudeney, *536 Puzzles and Curious Problems* (Souvenir Press, 1968). Edited.

    • H.E. Dudeney, *Puzzles and Curious Problems* (Fontana, 1970).
    • H.E. Dudeney, *More Puzzles and Curious Problems* (Fontana, 1970).

    Published in two volumes.

13. Kobon Fujimura, *The Tokyo Puzzles* (Muller, 1979*, 1981). Edited.
14. E. T. Bell, *Mathematics: Queen and Servant of Science* (Cambridge Univ Press, 1996*). Introduction by Gardner
15. Karl Menninger, *Calculator's Cunning* (Bell, 1964*). Gardner introduction.
16. Pierre Berloquin, *Geometric Games* (Unwin, 1980*). Gardner introduction.

    Also used in *Games of Logic,* (Unwin, 1982*) and perhaps others in the series.

17. Raymond Smullyan, *Alice in Puzzleland* (Penguin, 1984, 1987). Gardner introduction.
18. Rudy Rucker, *The Fourth Dimension and How to Get There* (Rider, 1985*). Gardner introduction.

- Rudy Rucker, *The Fourth Dimension and How to Get There* (Penguin, 1985*). Gardner introduction.

19. Jerry Slocum, Jack Botermans, *Puzzles Old & New* (Village Games, 1999). Gardner introduction.

Science

1. *Logic Machines and Diagrams: Second edition* (Harvester Press / Branch Line, 1983*).
2. *The Ambidextrous Universe* (Allen Lane, 1967).

   - *The Ambidextrous Universe* (Penguin, 1970*).

   Second edition published by Penguin, 1982, 1991*.

3. *The Sacred Beetle and Other Great Essays in Science* (Oxford University Press, 1985).

   - *Great Essays in Science* (Oxford University Press, 1997).

   Reissued "with revisions."

4. Roger Penrose, *The Emperor's New Mind* (Oxford University Press, 2002*). Gardner introduction.

Fringe Science

1. *Science: Good, Bad, and Bogus* (Oxford University Press, 1983*, 1984*).

Philosophy of Science

1. *Order & Surprise* (Oxford University Press, 1984).
2. *Gardner's Whys & Wherefores* (Oxford University Press, 1990).
3. *The Night is Large: Collected Essays 1938-1995* (Penguin, 1997).

Philosophy and Theology

1. *The Whys of a Philosophical Scrivener* (Harvester Press, 1983).

   - *The Whys of a Philosophical Scrivener* (Oxford University Press, 1985*).

Fiction

1. *The Flight of Peter Fromm* (Grafton, 1992*).

   "New edition"

   - *The Flight of Peter Fromm* (Paladin, 1999*).
2. *Visitors from Oz* (Penguin, 1999).

Literature

1. *The Annotated Ancient Mariner* (Anthony Blond, 1965*).
2. *The Annotated Innocence of Father Brown* (Oxford University Press, 1987).

   True first edition.

Carrolliana

1. *The Annotated Alice* (Anthony Blond, 1960*). "Manufactured" in the USA
   - *The Annotated Alice* (Penguin, 1965).

     An uncredited poem, in the shape of a tail, on the back cover is reprinted in *Knight Letter,* Spring 2020, p. 35.
   - *The Annotated Alice* [Revised edition] (Penguin, 1970).
   - *The Annotated Alice* (Nelson, 1975).

     First true British Edition, but from the same page images.
   - *The Annotated Alice–Definitive Edition* (Allen Lane, 2000).
   - *The Annotated Alice–Definitive Edition* (Penguin, 2001).
2. *The Annotated Snark* (Penguin, 1967, 1974).

   Revised edition with new introduction.

   - *The Hunting of the Snark* (Penguin, 1995).
     Title change for the Penguin Classics series.
3. *The Wasp in the Wig* (Macmillian, 1977).

Juvenile Literature

1. *Science Puzzlers* (Macmillan, 1962).
2. *Space Puzzles* (Bell, 1974*).

Miscellaneous

1. Anthony Ravielli, *Adventure in Shapes* (Phoenix House, 1967*).

   Retitling of *An Adventure in Geometry*, ghosted by Gardner

B.3   Bulgarian

Mathematics

1. Trans. Krasimira Popova [and Vladimir Sotirov in 1997], *Matematicheski Razvlecheniya: T. 1* (Sofia: Nauka, 1975, 1997). From *Mathematical Puzzles of Sam Loyd: Volume 1*
2. Trans. Krasimira Popova, *Matematicheski Razvlecheniya: T. 2* (Sofia: Nauka, 1977). From *Mathematical Puzzles of Sam Loyd: Volume 2*

B.4   Catalan

Oziana

1. Trans. Ramon Folch i Camarasa, *El Màgic d'Oz: Edició Anotada* (Barcelona: Editorial Empúries, 2002). Gardner introduction. From *The Annotated Wizard of Oz*, edited by M. P. O'Hearn

B.5   Chinese

Recreational Mathematics

1. Trans. [Lu Yueh-Sheng], [*Qu Wei Shu Xue*] [Interesting Mathematics] (Hong Kong: [Wan-Li Bookstore], 1977*). From *Mathematical Puzzles and Diversions*
2. Trans. [Feng Zongxin], [*Ke Xue Mei Guo Ren Qu Wei Shu Xue Ji Jin*] (Shanghai: [Shanghai Scientific and Technological Education Publishing], 2009). From *[First] Mathematical Puzzles and Diversions*
3. Trans. [Yi Wei-Wen], [*Mi gong, huang jin bi, suo ma li fang ti*] (Taipei: [Tian Xia Yuan Jian Chu Ban] Commonwealth Pub-

lishing Co. Ltd., 2003). From *Second Mathematical Puzzles and Diversions*

4. Trans. [Feng Zongxin], [*Mazes, Magic Squares, Geometrical Figures and Others*] (Shanghai: [Shanghai Scientific and Technological Education Publishing], 2008). From *Second Mathematical Puzzles and Diversions*

5. Trans. [Hu Shou-Ren], [*Ping Tu Ping Zi Ping Shu Xue*] [Puzzling with Pictures, Words and Mathematics] (Taipei: [Tian Xia Tuan Jian Chu Ban] Commonwealth Publishing Co. Ltd., 2005). From *New Mathematical Diversions*

6. Trans. [Ye Wei-Wen], [*Gui lun, pu ci zhuan, Bo luo mi ou huan*] (Taipei: [Tian Xia Tuan Jian Chu Ban] Commonwealth Publishing Co. Ltd., 2003). From *Unexpected Hanging and Other Mathematical Diversions*

7. Trans. [Xue Mei-zhen], [*A Ha! You Qu de Tui Li*] [Aha!, Interesting Reasoning] (Taipei: [Tian Xia Wen Hua] Commonwealth Publishing Co. Ltd., 1995*, 2001). in two volumes From *Aha! Insight*

8. Trans. [unknown], [*Ling Ji Miao Suan*] (Xinzhu Shi: [Fan Yi Ch Ban She], 1997). From *Aha! Insight*

9. Trans. [Hu Zuo Xuan], [*A Ha! Ling Ji Yi Dong*] [Aha!, Think up a Good Idea] (Beijing: Science Publishing, 2005). From *Aha! Insight*

10. Trans. [Cai Cheng-Zhi], [*Shu Xue Ma Xi Tuan*] [A Mathematical Circus Collection] (Taipei: Yuan-Liou Publishing Co. Ltd., 2005). From *Mathematical Circus*

11. Trans. [Yi Wei-Wen], [*Ge Lao Die De Tui Li You Xi*] Vols 1 and 2 (Taipei: Tian Xia Yuan Jian Chu Ban, 2002). From *Mathematical Puzzle Tales* and *Riddles of the Sphinx*

12. Trans. [Bai Baolin], [*Thoughts from Surprise*] (Shanghai: [Shanghai Scientific and Technological Education Publishing House], 1986). From *Paradox Box*

13. Trans. [Xue Mei-zhen], [*Tiao Chu Si Lu de Xian Jing*] [Spring Out from Pitfalls in Trains of Thought] (Taipei: [Tian Xia Wen Hua] Commonwealth Publishing Co. Ltd., 1988*, 1993*, 2001*). From *Aha! Gotcha*

14. Trans. [Xi Ma], [*Mao Dun Ji Jun*] [Paradoxes to Puzzle and Delight] (Xinzhu Shi: [Fan Yi Chu Ban She], 1996). From *Aha! Gotcha*

15. Trans. [Hu Zuoxuan], [*A Ha! Yuan Lai Ru Chi*] (Beijing: [Ke Xue Chu Ban She], 2008). From *Aha! Gotcha*

16. Trans. [Tan Xiang Bai], [*Ju Zhen Bo Shi De Mo Fa Shu*] (Shanghai: [Shanghai Scientific and Technological Education Publishing House], 2001). From *Magic Numbers of Dr Matrix*

17. Trans. [Tan Xiang Bai], [*Zui Hou De Xiao Qian*] [The Last Pastimes] (Shanghai: [Shanghai Scientific and Technological Education Publishing House], 2010). From *The Last Recreations*

18. Trans. [Hu Shou-Ren], [*Da Kai Mo Shu Xiang*] [Opening the Magic Number Box] (Taipei: Yuan-Liou Publishing Co. Ltd., 2004). From *A Gardner's Workout*

19. Trans. [Wei-Wen Ye], [*Ge Lao Die De Tai Li You Xi*] [Ge Laojiao's Reasoning Game] (Taipei: Tai bei Xian san chong shi, 2004). From *Mathematical Puzzle Tales* in 2 volumes

20. Trans. [Chen Peng], [*Lao Ai De De Shu Xue Qu Ti Xu Bian*] [Interesting Mathematical Puzzles] (Shanghai: [Shanghai Scientific and Technological Education Publishing House], 1999*, 2001*). From *Mathematical Puzzles of Sam Loyd*

21. Trans. [Tan Xiang Bo], [*Sa Mu – Lao Ai De De Shu Xue Qu Ti Xu Bian*] [Interesting Mathematical Puzzles – Sequel] (Shanghai: [Shanghai Scientific and Technological Education Publishing House], 1999*, 2001*). From *More Mathematical Puzzles of Sam Loyd*

22. Trans. [Chen Wei Peng], [*Qu Ti Da Sci De Tou Nao Ti Cao*] [Mind Gymnastics of the Puzzle Master] (Shanghai: [Shanghai Scientific and Technological Education Publishing House], 2010*). From *Mathematical Puzzles of Sam Loyd*

23. Trans. [Chen Wei Peng], [*Qu Ti Da Sci De Tou Nao Ti Cao*] [Mind Gymnastics of the Puzzle Master] (Shanghai: [Shanghai Scientific and Technological Education Publishing House], 2004*, 2010*). From *Mathematical Puzzles of Sam Loyd* in two volumes

Science

1. Trans. [Mai Lin], [*Da Zhong Xiang Dui Lun*] (Shanghai: [Shanghai Scientific and Technological Education Publishing House], 1989*). From *Relativity for the Million*, translated from a 1979 Russian edition

2. Trans. [Wang Qijun] and [Ke Xiuwen], [*Ke Xue Jia Yan Zhong De Shi Jie*] (Taipei: [Shang Zhou Chu Ban], 2005*). From *"The Sacred Beetle" and Other Great Essays in Science*

Fringe Science

1. Trans. [Yung-Hui Li], [*Aidisheng, Ni Bei Pian Le!*] [Edison, you have been fooled!] (Taiwan: Sino Cultural Enterprise / Rive Gauche Publishing House, 2004). From *Did Adam and Eve Have Navels?*
2. Trans. [Shen Liwen], [*Kan Kan Zhe Ge Bu Ke Xue De Yu Zhou*] [Let's look at this non-scientific universe] (Taipei: Yuan Liao Publ, 2006). From *Are Universes Thicker than Blackberries?*

Fiction

1. Trans. [Ruihua Huang], [*Yi ge xin tu de chu zou*] [One Believer's Exit] (Taipei: [Jiu Jing], 2002). From *The Flight of Peter Fromm*

Juvenile Literature

1. Trans. [Lin Zi-Xin], [*Yin Ren Ru Sheng de Shu Xue Qu Ti*] [Enchanting Mathematical Interesting Problems] ([Shanghai]: [Shanghai Scientific and Technological Education Publishing House], 1999). From *Entertaining Mathematical Puzzles*

B.6 Czech

Juvenile Literature

1. Trans. [unknown], *Zabavne Matematicke Hadanky* (Prague: Dokoran, 2018*). From *Mathematical Puzzles*

B.7 Danish

Recreational Mathematics

1. Trans. Carl-Otto Johansen, *Morsom Matematik* (Slagelse: Borgens Forlag, 1963). From *Mathematical Puzzles and Diversions*
2. Trans. Carl-Otto Johansen, *Mere Morsom Matematik* (Slagelse: Borgens Forlag, 1964). From *The Second Book of Mathematical Puzzles and Diversions*

Science

1. Trans. Gunvor Bjerre-Christensen, *Relativitetsteorierne* (København: Steen Hasselbalchs Forlag, 1964). From *Relativity for the Million*

B.8   Dutch

Recreational Mathematics

1. Trans. Bab Westerveld, *Magie en Mysterie met Mathematiek* (Amsterdam: Meulenhoff / Landshoff, 1982). From *Mathematics, Magic and Mystery*
2. Trans. Floor Gerard van Herwaarden, *Het Mathematische Circus* (Amsterdam: Uitgeverij Bert Bakker, 1983). From *Mathematical Circus*
3. Trans. Marinus Bier, *De Mathematische Kermis* (Amsterdam: Uitgeverij Contact, 1985). From *Mathematical Carnival*, 9 chapters
4. Trans. Patty Adelaar, *Het Mathematische Carnaval* (Amsterdam: Uitgeverij Contact, 1987). From *Mathematical Carnival*, 8 chapters
5. Trans. Bab Westerveld, *Sam Loyds Raadselboek* (Amsterdam: Muelenhoff / Landshoff, 1980*). From *Mathematical Puzzles of Sam Loyd: Vol 1*
6. Trans. Bab Westerveld, *Sam Loyds Tweede Raadselboek* (Amsterdam: Muelenhoff / Landshoff, 1981*). From *Mathematical Puzzles of Sam Loyd: Vol 2*

Science

1. Trans. O. Meijer, *Relativiteitstheorie voor Iedereen* (Utrecht: Prisma-Boeken, 1966). From *Relativity for the Million*
2. Trans. Georeg Beekman, *Spiegelsymmetrie: Links en Rechts in de Natuur* (Amsterdam: Aramith Uitgevers, 1986). From *The Ambidextrous Universe*

Fringe Science

1. Trans. H. M. Meelkop, *Is Dat Nog Wel Wetenschap?* (Utrecht: Prisma-Boeken, 1967). From *Fads and Fallacies*

Carrolliana

1. Trans. Nicolaas Matsier [aka Reinsma van Tjit], *De avonturen van Alice in Wonderland de Achter de Spiegel en wat Alice daar Aantrof Samen Met [Aantekeningen bij Alice]* (Amsterdam: Athenaeum-Polak & Van Gennep, 2009*). From *The Annotated Alice*

Juvenile Literature

1. Trans. J. W. F. Klein - Von Baumhauer, *100 Trucs: Goochelen met de Wetenschap* (Amsterdam: Elsevier, 1966). From *Science Puzzlers*

B.9 Estonian

Science

1. Trans. E.-R. Tammet, *Relatiivsusteooria Miljonitele* (Tallinn: Kirjastus Valgus, 1968). From *Relativity for the Million*, with a new foreword by A. Baz.
2. Trans. J. Lõhmus, *Parem-Vasak Maailm* (Tallinn: Kirjastus Valgus, 1972). From *The Ambidextrous Universe*, with a long afterword by the translator.

B.10 Finnish

Recreational Mathematics

1. Trans. Pertti Jotuni, *Älyniekka Jokamiehen Ongelmakitja* (Helsinki: Weilen and Göös, 1965). From chapters in the first and second books of *Mathematical Puzzles and Diversions*

   • *Älyniekka Jokamiehen Ongelmakitja* (Helsinki: Weilen and Göös, 1970).

   With an additional 150 puzzles by Kari J. Pekkanen

2. Trans. Antti Pietiläinen, *Ongelmatarinoita* (Helsinki: Terra Cognita, 2003). From *Mathematical Puzzle Tales*

B.11   French

Recreational Mathematics

1. Trans. R. Rosset, *Mathématiques Magie et Mystère* (Paris: Dunod, 1961). From *Mathematics, Magic and Mystery*
2. Trans. Richard Vollmer, *Mathématiques Magie et Mystère* (Strasbourg: Editions du Spectacle, 1991). From *Mathematics, Magic and Mystery*

   - *Mathématiques Magie et Mystère* (Strasbourg: Magix Unlimited, 1995).

3. Trans. R. Marchand, *Problèmes et Divertissements Mathématiques, Tome 1* (Paris: Dunod, 1964). From *Mathematical Puzzles and Diversions*; "Tome 1" only on later printings.
4. Trans. R. Marchand, *Problèmes et Divertissements Mathématiques, Tome 2* (Paris: Dunod, 1965). From *The Second Book of Mathematical Puzzles and Diversions*
5. Trans. Claude Roux, *Nouveaux Divertissements Mathématiques* (Paris: Dunod, 1970). From *New Mathematical Diversions*
6. Trans. Claude Roux, *Le Paradoxe du Pendu et Autres Divertissements Mathématiques* (Paris: Dunod, 1971). From *The Unexpected Hanging*
7. Trans. Y. Roussel, *Jeux Mathématiques du 'Scientific American'* (Paris: Cedic, 1979*). From *The Sixth Book of Mathematical Games*

   - *Jeux Mathématiques du 'Scientific American'* (Amiens: Association pour le développement de la culture scientifique (ADCS), 1996*). APMEP (l'Association des Professeurs de Mathématiques de l'Enseignement Public) brochure number 402.

8. Trans. Pierre Tougne, *Jeux Mathématiques de 'Pour la Science'* (Paris: Bibliothèque pour la Science, 1980*). From columns rather than a book
9. Trans. Pierre Tougne, *La Mathématiques de Jeux* (Paris: Belin / Bibliothèque pour la Science, 1977*, 1991*). From columns rather than a book, "18 articles sont parus dans *Pour la science* de 1977 'a 1990."
10. Trans. [Not given], *Math' Festival* (Paris: Bibliothèque pour la Science, 1981). From *Mathematical Magic Show*
11. Trans. Jeanne Peiffer, *Haha: Ou L'éclair de la Compréhension Mathématique* (Paris: Bibliothèque pour la Science, 1979, 1992*). From *Aha! Insight*

12. Trans. Jean-Pierre Labrique, *Math' Circus* (Paris: Bibliothèque pour la Science, 1982, 1986*). From *Mathematical Circus*

13. Trans. Agnès Szakonyi, *Casse-Tête Dans le Cosmos* (Paris: Dunod, 1982). From *Science Fiction Puzzle Tales*

14. Trans. [Not given], *La Magie des Paradoxes* (Paris: Bibliothèque pour la Science, 1980, 1985*, 199, 1997). From *Paradox Box* which became *Aha! Gotcha* in 1982

15. Trans. [Bernard Malgrange], *Le Monde Mathématique* (Paris: Bibliothèque pour la Science / Belin, 1984*, 1986*).

    With a new introduction by the translator(?)

16. Trans. F. Rostas, *Les Casse-Tête Mathématiques de Sam Loyd, Tome 1* (Paris: Dunod, 1964). From *Mathematical Puzzles of Sam Loyd*, Volume 1

17. Trans. Ph. Gatbois, *Les Casse-Tête Mathématiques de Sam Loyd, Tome 2* (Paris: Dunod, 1967). From *Mathematical Puzzles of Sam Loyd*, Volume 2

18. Trans. F. Rostas and Ph. Gatbois, *Les Casse-Tête Mathématiques de Sam Loyd* (Paris: Dunod, 1970*, 1979*). In one volume. From *Mathematical Puzzles of Sam Loyd*

    With a preface by Pierre Berloquin

19. Trans. Christian Jeanmougin [and later with David Povilaitis], *La Quatrième Dimension* (Paris: Seuil, 1985*, 2001*). Gardner introduction. From *The Fourth Dimension*, by Rudy Rucker

Science

1. Trans. Mary Ghezzi, *L'Étonnante Histoire des Machines Logiques* (Paris: Dunod, 1964). From *Logic Machines and Diagrams*

2. Trans. A. Gosset, *La Relativité pour Tous* (Paris: Dunod, 1969). From *Relativity for the Million*

3. Trans. Claude Roux, *L'Univers Ambidextre: La Droite, la Gauche et la Faillite de la Parité* (Paris: Dunod, 1968*). From *The Ambidextrous Universe*

   - *L'Univers Ambidextre: Les Miroirs de L'Espace-Temps* (Paris: Éditions de Seuil, 1985, 1994*).

     A revised edition, including changes that had not yet appeared in English.

   - Trans. Claude Roux and Alain Laverne, *L'Univers Ambidextre: Les Symétries de la Nature* (Paris: Éditions de Seuil, 2000*). From Revised edition

4. Trans. Francoise Balibar and Claudine Tiercelin, *Computerdenken* (Paris: InterEditions, 1992). Gardner introduction. From *The Emperor's New Mind*, by Roger Penrose

Fringe Science

1. Trans. Béatrice Rochereau, *Les Magiciens Démasqués: Santé et Prospérité des Pseudo-Savants* (Paris: Presses de la Cité, 1966). From *Fads and Fallacies*

Philosophy of Science

1. Trans. Jean-Mathiea Luccioni and Antonia Soulez, *Les Fondements Philosophiques de la Physique* (Paris: Armand Colin, 1973). From *Philosophical Foundations of Physics*, by Rudolph Carnap

Literature

1. Trans. Emile Cammaerts, *La Clairvoyance du Père Brown* (Paris: Signe de Piste Editions, 1992*). From *The Annotated Innocence of Father Brown*, G K Chesterton.

Carrolliana

1. Trans. Bruno Roy, *Le Frelon a Perruque: Un Chapitre Inédit de L'autre côté de Miroir* (Montpellier: Bibliothèque Artistique et Littéraire, 1979*). Gardner annotation. From *The Wasp in the Wig*, by Lewis Carroll; illustrated by Anne-Marie Soulcié.

Juvenile Literature

1. Trans. Claudine Azoulay, *Remue-Méninges Captivants* (Montreal: Editions Bravo, 2009*). From *Classic Brainteasers*
2. Trans. unknown, *Le Grand Cirque des Illusions d'Optique* (Paris: Fleurus, 2009*). From *Optical Illusion Play Pack*

Magic

1. Trans. Alain Midan, *Encyclopédie de la Magie Impromptue: Tome 1* (Montpellier: Passe-Passe, 2002*). Also listed as from

Joker Deluxe. From *The The Encyclopedia of Impromptu Magic*, adapted by the translator and François Montmirel

2. Trans. Alain Midan, *Encyclopédie de la Magie Impromptue: Tome 2* (Montpellier: Passe-Passe, 2004*). Also listed as from Joker Deluxe

B.12   German

Recreational Mathematics

1. Trans. Matthias Schramm, *Mathematik und Magie* (Köln: Du-Mont Buchverlag, 1981, 1986*, 2002). From *Mathematics, Magic and Mystery*, with a new foreword by Alexander Adrion

   - *Mathematische Zaubereien* (Köln: DuMont Literatur und Kunst Verlag, 2004).
   - *Das Verschwundene Kaninchen* (Köln: DuMont Literatur und Kunst Verlag, 2016).

2. Trans. Brigitte Kunisch, *Mathemagische Tricks* (Braunschweig: Vieweg, 1981). From *Mathematics, Magic and Mystery*

3. Trans. Patrick P. Weidhaas, *Mathematische Rätsel und Probleme* (Braunschweig: Vieweg, 1966, 1971*, 1975, 1990*). From Selections from the first two *Mathematical Puzzles and Diversions*, with a new foreword by Roland Sprague.

4. Trans. Eberhard Bubser, *Mathematische Knobeleien* (Braunschweig: Vieweg, 1973, 1984, 2013*). From *New Mathematical Diversions*

5. Trans. Astrid Gorvin, *Die Zahlenspiele des Dr. Matrix* (Frankfurt/M: Ullstein Ratgeber, 1981). From *The Numerology of Dr. Matrix*

6. Trans. Carlo Karrenbauer, *Logik unterm Galgen* (Braunshweig: Vieweg, 1971, 1978*, 1980*, 1982*, 2013*). From *The Unexpected Hanging*

7. Trans. [Not given], *Kopf oder Zahl? Paradoxa und Mathematische Knobeleien* (Weinheim: Spektrum der Wissenschaft, 1978). From a small booklet contain some chapters from *Mathematische Knobeleien* and *Logik unter dem Galgen* given as premium to subscribers of the German edition of *Scientific American*.

8. Trans. Brigitte Kunisch and Rudolf Heersink, *Mathematisches Labyrinth* (Braunschweig: Vieweg, 1979). From *Sixth Book of Mathematical Games*

9. Trans. Sibylle and Ulfreid Wesser, *Mathematischer Karneval* (Frankfurt/M: Ullstein, 1975*, 1977, 1980, 1986*, 1993*, 1999*). From *Mathematical Carnival*

10. Trans. Brigitte Mitchel and Gerd Bartmann, *Mathematische Hexereien* (Frankfurt/M: Ullstein, 1979, 1988*). From *Mathematical Magic Show*, two covers both dated 1979

11. Trans. Immo Diener and Winfried Petri, *Aha! Oder das Wahre Verständnis der Mathematik* (Heidelberg: Spektrum, 1981). From *Aha! Insight*

    - *Aha! Oder das Wahre Verständnis der Mathematik* (München: Hugendubel, 1984, 1987*).
    - *Aha! Oder das Wahre Verständnis der Mathematik* (München: Deutscher Taschenbuch-Verlag, 1986).

12. Trans. Reinhard Soppa, *Mathematicher Zirkus* (Frankfurt/M: Ullstein, 1984, 1988*). From *Mathematical Circus*, two covers dated 1984

13. Trans. Peter Ripota, *Denkspiele aus der Zukunft* (München: Hugendubel, 1982, 1985*, 1991*). From *Science Fiction Puzzle Tales*

    - *Denkspiele aus der Zukunft* (München: Büchergilde Gutenberg, 1983*).

14. Trans. Irene Rumler, *Gotcha: Paradoxien für den Homo Ludens* (München: Hugendubel, 1985). From *Aha! Gotcha: Paradoxes to Puzzle and Delight*

    - *Gotcha: Paradoxien für den Homo Ludens* (München: Deutscher Taschbuch Verlag, 1987).

15. Trans. Peter Ripota, *Mathematische Denkspiele* (MüChen: Hugendubel, 1987, 1990*). From *Wheels, Life, and Other Mathematical Amusements*

16. Trans. Peter Ripota, *Denkspiele von Anderen Planeten* (München: Hugendubel, 1986, 1990*). From *Puzzles from Other Worlds*

17. Trans. Thea Brandt and Gerhard Trageser, *Die Magischen Zahlen des Dr. Matrix* (Frankfurt/M: Wolfgang Krüger Verlag / Fischer, 1987, 1990*). From *The Magic Numbers of Dr. Matrix*

18. Trans. Klaus Volkert, *Bacons Geheimnis* (Frankfurt/M: Wolfgang Krüger Verlag, 1986*, 1990). From Selections from *Knotted Doughnuts and Other Mathematical Entertainments*

19. Trans. Klaus Volkert, *Mit dem Fahrstuhl in die 4 Dimension* (Frankfurt/M: Wolfgang Krüger Verlag, 1991). From Selections from *Knotted Doughnuts and Other Mathematical Entertainments* and *Time Travel and Other Mathematical Bewilderments*

20. Trans. Anita Ehlers, *Geometrie mit Taxis, die Köpfe der Hydra und andere Mathematische Spielereien* (Basel: Birkhäuser, 1997). From *The Last Recreations*

21. Trans. Charlotte Franke, *Mathematische Rätsel and Spiele: Denksportaufgaben für Kluge Köpfe: 117 Aufgaben und Lösengen* (Köln: DuMont, 1978*, 1995*, 2002*, 2006*, 2012*). Edited. From *Mathematical Puzzles of Sam Loyd*, Volume 1

22. Trans. Charlotte Franke, *Noch Mehr Mathematische Rätsel and Spiele: Denksportaufgaben für Kluge Köpfe: 166 Aufgaben und Lösengen* (Köln: DuMont, 1979, 1991*, 2002*). Edited. From *Mathematical Puzzles of Sam Loyd*, Volume 2

23. Trans. Charlotte Franke, *Vom Kken Zum Ei: Noch Mehr Mathematische Rätsel and Spiele* (Köln: DuMont, 2015*). Edited. From *Mathematical Puzzles of Sam Loyd*, Volume 2, retitled

24. Trans. Charlotte Franke, *Noch Mehr Mathematische Rätsel and Spiele: Denksportaufgaben für Kluge Köpfe: 283 Aufgaben und Lösengen* (Köln: DuMont, 1979, 2002*, 2005*, 2008*). Edited. From *Mathematical Puzzles of Sam Loyd* in one volume

25. Trans. –, *Alice im Rätselland* (Fankfurt am Main: Fischer, 1988). Gardner introduction. From *Alice in Puzzle-Land* by Raymond Smullyan

Science

1. Trans. Charles Hummel, *Relativitätstheorie für Alle* (Zürich: Orell Füssli, 1966*, 1969). From *Relativity for the Million*

   • Trans. Charles Hummel and Eva Sobottka, *Relativitätstheorie für Alle* (Köln: DuMont, 2005). From *Relativity Simply Explained*

2. Trans. Winfried Petri and Carlo Karrenbauer, *Das Gespiegelte Universum* (Braunschweig: Vieweg, 1967, 1982*). From *The Ambidextrous Universe*, contains a new "Vorwort" by Petri, correcting two errors.

3. Trans. Gerd Bartmann, *Unsere Gespiegelte Welt: Denksportaufgaben und Zaubertricks* (Berlin: Ullstein, 1982, 1987*). From *The Ambidextrous Universe*, second edition

4. Trans. Michael Springer, *Computerdenken* (Heidelberg: Spektrum Akademischer, 1991). Gardner introduction. From *The Emperor's New Mind*, by Roger Penrose

Fringe Science

1. Trans. Gerd Bartmann, *Kabarett der Täuschungen: Unter dem Deckmantel der Wissenschaft* (Berlin: Ullstein, 1983). From *Science: Good, Bad, and Bogus*

Philosophy of Science

1. Trans. Walter Hoering, *Einführung in die Philosophie der Naturwissenschaft* (München: Nymphenburger, 1969, 1974, 1986). Edited. From *Philosophical Foundations of Physics*, by Rudolph Carnap

Carrolliana

1. Trans. Flemming, Günther, *Alles über Alice* (Hamburg: Europa Verlag, 2002). From *The Annotated Alice: The Definitive Edition*, with additional translation by Friedhelm Rathjen

Juvenile Literature

1. Trans. Sarah Kielich and Brigitte Michel, *Rätsel und Denkspiele* (Frankfurt/M: Ullstein Ratgeber, 1981). From *Perplexing Puzzles and Tanatalizing Teasers* and *More Perplexing Puzzles and Tanatalizing Teasers*, in one edition.
2. Trans. Astrid Gorvin, *Das Verhexte Alphabet: Tips und Tricks für Geheimschriften* (Frankfurt/M: Ullstein Ratgeber, 1981). From *Codes, Ciphers, and Secret Writing*
3. Trans. Reinhard Soppa, *Mathematische Planetenzauberei* (Berlin: Ullstein, 1980). From *Space Puzzles*

B.13   Greek

Recreational Mathematics

1. Trans. [Th. Papadópoulos], [*To Panēgýri tōn Mathēmatikōn*] [The Feast of Mathematics] ([Athens]: [Trochalía], 1986). From *Mathematical Carnival*

2. Trans. [Gavras Kostas], [*To Ainigmata tōn Sfiggas*] [The Enigma of Sphinx] ([Katoptro], 1998). From *Riddles of the Sphinx*

3. Trans. [O. Papadópoulos], [*o Tsirko tōn Mathēmatikōn*] [The Circus of Mathematics] ([Athens]: [Trochalía], 1990). From *Mathematical Circus*

4. Trans. [Tasoúla Dimitríou and Gregórēs Troufákos], [*Ē Mageía tōn Paradóxōn*] [The Magic of Paradoxes] ([Athens]: [Trochalía], 1989). From *Aha! Gotcha*

B.14   Hebrew

Mathematics

1. Trans. David Poviliatis, *ha-Memad ha-Revi'i* ([Jerusaleum]: Hotso'at Sefarim 'a sh. Y.L. Magnes, ha-Universitah ha-'Ivrit, 2000). Gardner introduction. From *Fourth Dimension* by Rudy Rucker

Carrolliana

1. Trans. Rina Litvin, [*Alice in Wonderland (The New Library No. 13)*] ([Tel Aviv]: [United Kibbutz Publishing House], 1997). From *Annotated Alice* and *More Annotated Alice* (but only *Alice in Wonderland*) with additional material by the translator.

2. Trans. Rina Litvin, [*Through the Looking Glass*] ([Tel Aviv]: [United Kibbutz Publishing House], 1999). From *Annotated Alice* and *More Annotated Alice* (but only *Through the Looking Glass*) with additional material by the translator. Also contains *The Wasp in the Wig*.

B.15   Icelandic

Recreational Mathematics

1. Trans. Benedikt Jóhannesson, *Ekkï Er Allt Sem Sýnist* (Reykjavík: Talnakönnun, 1989). From *Aha! Gotcha*, with a new foreword by the translator.

B.16   Italian

Recreational Mathematics

1. Trans. Carroll Mortera, *I Misteri della Magia Matematica* (Firenze: Sansoni, 1985). From *Mathematics, Magic and Mystery*, with a new introduction by Ennio Peres
2. Trans. Mario Carlà, *Enigmi e Giochi Matematici* ([Firenze]: Sansoni, 1967, 1972, 1983*, 1990*). From *Mathematical Puzzles and Diversions*

   - Trans. Mario Carlà, *Enigmi e Giochi Matematici* (Milan: Biblioteca Universale Rizzoli, 1987*, 1998*, 2001*, 2005, 2010).

3. Trans. Virginio B. Sala, *Come Buttarsi Dalla Torre di Hanoi: Enigmi e Giochi Matematici* (Firenze: Maria Margherita Bulgarini / Emmebi Edizioni, 2012). From *Hexaflexagons, Probability Paradoxes, and the Tower of Hanoi*
4. Trans. Mario Carlà, *Enigmi e Giochi Matematici - 2* ([Firenze]: Sansoni, 1968, 1972*, 1973, 1983*, 1990*). From *The Second Book Mathematical Puzzles and Diversions*
5. Trans. Mario Carlà, *Enigmi e Giochi Matematici: Indovinelli, problemi, paradossi per provare la vostra intelligenza* (Milano: Biblioteca Universale Rizzoli, 1987, 1997*, 2000*). From Combined edition of the first two *Mathematical Puzzles and Diversions*
6. Trans. Mario Carlà, *Enigmi e Giochi Matematici - 3* (Firenze: Sansoni, 1969, 1973, 1990*). From *New Mathematical Diversions*
7. Trans. Mario Carlà, *Enigmi e Giochi Matematici - 4* (Firenze: Sansoni, 1975, 1977, 1985*, 1990*). From *The Unexpected Hanging*
8. Trans. Mario Carlà, *Enigmi e Giochi Matematici - 5* (Firenze: Sansoni, 1976, 1985*, 1990*). From *Sixth Book of Mathematical Games*
9. Trans. Bianca Rosa Bellomo Bove, *Carnevale Matematico* (Bologna: Zanichelli, 1977, 1980*). From *Mathematical Carnival*

   - *Carnevale Matematico* (Bologna: RBA Italia, 2008).

10. Trans. Simona Panattoni, *L'Incredible Dottor Matrix* (Bologna: Zanichelli, 1982). From *The Incredible Dr. Matrix*
11. Trans. Simona Panattoni, *Show di Magia Matematica* (Bologna: Zanichelli, 1980). From *Mathematical Magic Show*

    - *Show di Magia Matematica* (Bologna: RBA Italia, 2008).

12. Trans. Angela Iorio, *Esperienza A-Ah!* (Milano: RBA Italia, 2008). From *Aha! Insight*
13. Trans. Silvia Bemporad, *Circo Matematico* (Firenze: Sansoni, 1981). From *Mathematical Circus*
14. Trans. Simona Panattoni, *Ah! Ci Sono! Paradossi Stimolanti e Divertenti* (Bologna: Zanichelli, 1987). From *Aha! Gotcha*

   - *Ah! Ci Sono! Paradossi Stimolanti e Divertenti* (Novara: Mondadori-De Agostini, 1995*).
   - *Ah! Ci Sono! Paradossi Stimolanti e Divertenti* (Bologna: RBA Italia, 2008).

15. Trans. Angela Ioria and Rossella Pederzoli , *Viaggio nel Tempo e Altre Stranezze Matematiche* (Bologna: RBA Italia, 2008). From *Time Travel*
16. Trans. Silvia Caldara, *Enigmi da Altri Mondi: Fantastici rompicapo dalla rivista di fantascienza di Isaac Asimov* (Firenze: Sansoni, 1986). From *Puzzles from Other Worlds*, with a new introduction by Ennio Peres.
17. Trans. Roberto Morassi, *Passatempi Matematici, Vol. 1* (Firenze: Sansoni, 1980). From *Mathematical Puzzles of Sam Loyd*, Vol. 1
18. Trans. Roberto Morassi, *Passatempi Matematici, Vol. 2* (Firenze: Sansoni, 1980). From *Mathematical Puzzles of Sam Loyd*, Vol. 2
19. Trans. Silvia Mori, *Giochi Matematici Russi* (Firenze: Sansoni, 1982). From *The Moscow Puzzles*, by Boris Kordemsky

Science

1. Trans. Gianluca Pompei, *Nel Nome della Scienza* (Ancona: Transeuropa, 1998*). From *Fads and Fallacies*
2. Trans. Giovanna Martini Albini, *La Relatività per Tutti* (Firenze: Sansoni, 1965). From *Relativity for the Million*

   - *Che Cos' è la Relatività* (Firenze: Sansoni, 1977).

3. Trans. Luca Peranza and Guido Pedrazzini, *L' Universo Ambidestro* (Bologna: Zanichelli, 1984). From *The Ambidextrous Universe*

   - *L' Universo Ambidestro* (Novara: Mondadori-De Agostini, 1996*).

4. Trans. Libero Sosio, *La Mente Nuova dell'Imperatore* (Milan: Rizzoli, 1992*). From *The Emperor's New Mind*, by Roger Penrose; introduction by Gardner

Fringe Science

1. Trans. [Carlo Boriello], *Scienza, Imposture e Abbagli* (Milan: Ulrico Hoepli, 2006). From *Are Universes Thicker than Blackberries?*
2. Trans. [Federico Tibone], *Dracula, Platone e Darwin:Giochi matematici e riflessioni sul mondo* (Bologna: Zanichelli, 2010). From *The Jinn from Hyperspace*
3. Trans. [Fara Di Maio], *Confessioni di un Medium* (Padova: CICAP, 2006). Number 1 in the series *I Quaderni di Magia* From *Confessions of a Psychic* and *Further Confessions of a Psychic*

Philosophy of Science

1. Trans. Corrado Mangione and Emanuele Vinassa de Regny, *I Fondamenti Filosofici della Fisica* (Milano: il Saggiatore, 1971, 1982*). From *Philosophical Foundations of Physics*, by Rudolph Carnap

Carrolliana

1. Trans. Masolino d' Amico, *Alice: Le Avventure di Alice Nel Paese Delle Meravigile & Attraverso Lo Specchio E Quello Che Alice Vi Trovò* (Milano: Longanesi, 1971, 1984*). From *The Annotated Alice*
2. Trans. Masolino d' Amico, *Alice Nel Paese Delle Meravigile & Attraverso Lo Specchio E Quello Che Alice Vi Trovò* (Milano: Biblioteca Univ. Rizzoli, 2010*, 2015 (2016 with movie tie-in obi)). From *The Annotated Alice: Definitive Edition*

Juvenile Literature

1. Trans. Manlio Guardo, *Indovinelli Nello Spazio* (Bologna: Zanichelli, 1972*, 1979). From *Space Puzzles*

B.17 Japanese

Recreational Mathematics

1. Trans. [Yashinau Kanazawa], [*Sugaku Majikku*] [Mathematical Magic] ([Chiyoda-ku, Tokyo]: [Hakuyosha], 1959, 1999). From *Mathematics, Magic and Mystery*

2. Trans. [Yashinau Kanazawa], [*Gendai no Goraku Sugaku – Atarashii Pazuru, Majikku, Gehmu*] [Modern Mathematical Recreations – New Puzzles, Magics, and Games] ([Chiyoda-ku, Tokyo]: [Hakuyosha], 1960). From *Mathematical Puzzles and Diversions*

3. Trans. [Shin Hitotsumatsu], [*Sugaku Gemu*] [Mathematical Puzzles] (Nikkei Saiensu Sha, 1979*, 1980*, 1981*, 1982*). From *Mathematical Puzzles and Diversions* and the *Second Book*, in four volumes over four years.

4. Trans. [Yashinau Kanazawa], [*Omoshiroi Sugaku Pazuru I*] [Amusing Mathematical Puzzles I] ([Shakai Shisou Sha], 1980). From *Mathematical Puzzles and Diversions*, in two volumes.

5. Trans. [Hirokazu Iwasawa and Ryuhei Uehara], [*Furekusagon Kakuritsu Paradokkusu Poriomino*] (Tokyo: [Nihon Hyoronsha], 2015). From *Hexaflexagons, Probability Paradoxes and the Tower of Hanoi*

6. Trans. [Yashinau Kanazawa], [*Atarashii Sugaku Gemu-Pazuru*] [New Mathematical Game-Puzzle] ([Chiyoda-ku, Tokyo]: [Hakuyosha], 1966). From *The Second Book of Mathematical Puzzles and Diversions*

7. Trans. [Yashinau Kanazawa], [*Omoshiroi Sugaku Pazuru II*] [Amusing Mathematical Puzzles II] ([Shakai Shisou Sha], 1981). From *The Second Book of Mathematical Puzzles and Diversions*

8. Trans. [Ryuhei Uehara and Hirokazu Iwasawa], [*Soma Kyubu Ereushisu Seihokei no Seiho Gunkatsu*] (Tokyo: [Nihon Hyoronsha], 2015). From *Origami, Eleusis and the Soma Cube*

9. Trans. [Shigeo Takagi], [*Sugaku Gemu*] [Mathematical Games I and II] ([Bunkyo, Tokyo]: [Kodansha], 1974). From *The Unexpected Hanging*, in two volumes.

10. Trans. [Ryuhei Uehara and Hirokazu Iwasawa], [*Kyu o Tsumekomu Yonshoku Teiri Sabaunho*] (Tokyo: [Shakai Shisou Sha], 2015). From *Sphere Packing, Lewis Carroll and Reversi*

11. Trans. [Shin Hitotsumatsu], [*Sugaku Kahnibaru*] [Mathematical Carnival] ([Shinjuku, Tokyo]: [Kinokuniya Shoten], 1977). From *Mathematical Carnival*, in two volumes.

12. Trans. [Shin Hitotsumatsu], [*Meitorikkusu hakushi no koui no sunhijutsu*] [Dr. Matrix's Marvelous Numerology] ([Shinjuku]: [Kinokuniya Shoten], 1978). From *The Incredible Dr. Matrix*

13. Trans. [Shin Hitotsumatsu], [*Sugaku Mahokan*] [Gardner's Mathematical Magic Hall] ([Bunkyo, Tokyo]: [Tokyo Tosho], 1979). From *Mathematical Magic Show*, in three volumes.

14. Trans. [Kazou Shimada], [*Aha! – Hirameki Shikou*] [Aha! Insight Thinking] ([Chiyoda-ku, Tokyo]: [Nippon Keizai Shimbunsha], 1979). From *Aha! Insight*

   - [*Aha! – Hirameki Shikou*] [Aha! Insight Thinking] (Tokyo: Nikkei Saiensu Sha, 1983).
   - [*Aha! – Hirameki Shikou*] [Aha! Insight Thinking] (Tokyo: [Nippon Keizai Shimbunsha], 2009). in two volumes

15. Trans. [Hiroshi Takayama], [*Sugaku Sakasu*] [Mathematical Circus] ([Bunkyo, Tokyo]: [Tokyo Tosho], 1981). From *Mathematical Circus*

16. Trans. [Kenkichi Uejima], [*SF Pazuru*] [SF Puzzles] ([Shinjuku, Tokyo]: [Kinokuniya Shoten], 1982). From *Science Fiction Puzzle Tales*

17. Trans. [Akihiro Nozaki], [*The Paradox Box – Gyakusetsu no Shikou*] [The Paradox Box – The Thinking of Paradox] ([Chiyoda-ku, Tokyo]: [Nippon Keizai Shimbunsha], 1979). From *Paradox Box*

18. Trans. [Ikuo Takeuchi], *Aha Gotcha: [Yukaina Paradokkusu]* [Joyful Paradox] (Tokyo: Nikkei Saiensu Sha, 1982, 1995*). From *Aha! Gotcha*, in two volumes

   - *Aha Gotcha: [Yukaina Paradokkusu]* (Tokyo: [Nippon Keizai Shimbunsha], 2009). in two volumes

19. Trans. [Shin Hitotsumatsu], [*Arisutoteresu no Wa to Kauritsu no Sakkaku*] [Aristotle's Wheel and the Illusion of Probability] ([Chiyoda-ku, Tokyo]: [Nippon Keizai Shimbunsha], 1993). From *Wheels, Life and Other Mathematical Amusements*

20. Trans. [Kojiro Kuroda], [*Sufinkusu no nazo*] [Mystery of the Sphinx] ([Chiyoda-ku, Tokyo]: [Maruzen], 1989). From *Riddles of the Sphinx*

21. Trans. [Shin Hitotsumatsu], [*Penrozu Tairu to Sugaku Pazuru*] [Penrose Problem and Mathematical Puzzles] and [*Otoshido Angou no Nazotoki*] [Solving Riddles of Trapdoor Ciphers] and [*Meitorikkusu Hakase no Seikan*] [Return of Dr. Matrix] ([Chiyoda-ku, Tokyo]: [Maruzen], 1992). From *Penrose Tiles and Trapdoor Ciphers*, in three volumes.

22. Trans. [Shin Hitotsumatsu], [*Furakutaru Ongaku*] [Fractal Music] and [*Chonoryoku to Kakuritsu*] [Supernatural Power and Probability] and [*Enshuritsu to Shi*] [Pi and Poetry] ([Chiyoda-ku, Tokyo]: [Maruzen], 1996). From *Fractal Music, Hypercards and More*, in three volumes.

23. Trans. [Abe Takehisa, Idogawa Tomoyuki, and Fujii Yasuo], [*Gadona Kessaka Senshu*] ([Tokyo]: [Morikita], 2009). From *A Gardner's Workout*

24. Trans. [Shin Hitotsumatsu], [*Math Games I: New Edition of Martin Gardner*] ([Nikkei Science], 2010). From A new collection of *Scientific American* chapters. "[Supplement 176]"

25. Trans. [Shin Hitotsumatsu], [*Math Games II: New Edition of Martin Gardner*] ([Nikkei Science], 2011). From A new collection of *Scientific American* chapters. "[Supplement 182]"

26. Trans. [Shin Hitotsumatsu], [*Math Games III: New Edition of Martin Gardner*] ([Nikkei Science], 2013). From A new collection of *Scientific American* chapters. "[Supplement 190]"

27. Trans. [Kozaburo Fujimura], [*H. E. Dudeney's Pazuru Kessaku Shuu*] [H. E. Dudeney's Masterpieces of Puzzles] ([Chiyoda-ku, Tokyo]: [Kawade Shobo Shinsha], 1969). Edited. From *536 Puzzles and Curious Problems*, by H. E. Dudeney

28. Trans. [Isamu Tanaka], [*Samu Roido no Sugaku Paruzu*] [Sam Loyd's Mathematical Puzzles] ([Chiyoda-ku, Tokyo]: [Hakuyosha], 1965). Edited From *Mathematical Puzzles of Sam Loyd*, in three volumes. Also issued as [*Samu Roido no Paruzu Hyakka*] (1966).

    • Trans. [Isamu Tanaka], [Sam Loyd's Puzzle Encyclopedia] ([Chiyoda-ku, Tokyo]: [Hakuyosha], 1986*, 2000*). From *Mathematical Puzzles of Sam Loyd*, in three volumes.

29. Trans. [Monma Yoshiyuki and Monma Naoko], [*Lewis Carroll's Asobi no Ychu*] [Lewis Carroll's Recreations of the Universe] ([Chiyoda-ku, Tokyo]: Hakuyosha, 1998). From *The Universe in a Handkerchief*

30. Trans. [Yasuo Ichiba], [*Pazuru Rando No Arisu*] (Tokyo: [Shakaishisosha], 1985). Gardner introduction. From *Alice in Puzzle-Land* by Raymond Smullyan

Science

1. Trans. [Tsutomu Kaneko], [*Hyakuman-nin no Soutaiseiriron*] [Theory of Relativity for Million] ([Chiyoda, Tokyo]: [Hakuyosha], 1966). From *Relativity for the Million*

2. Trans. [Tsutomu Kaneko], [*Soutaiseiriron ga Kouiteki ni Yoku Wakaru Hou*] [The Book by Which We Understand Relativity] ([Chiyoda-ku, Tokyo]: [Hakuyosha], 1992). From *The Relativity Explosion*

  - [*Gadona no Sotaisei Riron Nyumon*] [Gardner's Theory of Relativity Introduction] ([Chiyoda-ku, Tokyo]: [Hakuyosha], 2007). From *The Relativity Explosion*

3. Trans. [Chuji Tsuboi and Hiroshi Kojima], [*Shizenkai ni Okeru Hidari to Migi*] [Left and Right in Nature] ([Shinjuku, Tokyo]: [Kinokuniya-Shoten], 1971). From *The Ambidextrous Universe*

4. Trans. [Chuji Tsuboi, Akihiko Fujii, and Hiroshi Kojima], [*Shinpan Shizenkai ni Okeru Hidari to Migi*] [Left and Right in Nature (New Ed.)] ([Shinjuku, Tokyo]: [Kinokuniya Shoten], 1992). From *The New Ambidextrous Universe*

Fringe Science

1. Trans. [Yasuo Ichiba], [*Kimyo na Ronri: Damasare yasusu no Kenkyu*] [Strange Logic: Research Easiness] (Tokyo: Shakai Shisosha [Social Thought], 1980, 1989, 2002*). From *In the Name of Science*

2. Trans. [Yasuo Ichiba], [*Kimyo na Ronri*] [Strange Logic II: Uri Geller to the Flying Saucer] (Tokyo: Shakai Shisosha [Social Thought], 1980, 1989*, 1992*). From *In the Name of Science*

3. Trans. [Yasuo Ichiba], *Kimyona Ronri* [Strange Logic I: Studies in Easiness] (Tokyo: Hayakawa, 2003). From *In the Name of Science*, first of two volumes

4. Trans. [Yasuo Ichiba], [*Strange Logic II: Why be Attracted to False Science*] (Tokyo: Hayakawa, 2003). From *In the Name of Science*, second of two volumes

5. Trans. [Jiro Ito], [*Inchiki Kagaku no Kaidokuho: Tsuitsui Shinjite Shimau Tondemo Gakusetsu*] (Tokyo: [Kobunsha] Japan-UNI, 2004). From *Did Adam and Eve have Navels?*

Philosophy of Science

1. Trans. [Mitsushige Sawada, Kojiro Nakayama and Eturo Mochimaru], [*Buturigaku no Tetsugakuteki Kiso*] [Philosophical Foundations of Physics] ([Chiyoda-ku, Tokyo]: [Iwanami Shoten], 1968).

Boxed. From *Philosophical Foundations of Physics*, by Rudolph Carnap

Carrolliana

1. Trans. Naoke Yankee, [*Katsura o Kabutta Suzumebachi*] ([Tokyo]: [Rengashobo Shin Sha], 1978). From *The Wasp in the Wig*
2. Trans. [Hiroshi Takayama], [*Fushigi no Kunino Arisu*] [Annotated Alice's Adventures in Wonderland] and [*Kagami no Kunino Arisu*] [Annotated Through the Looking Glass] ([Bunkyo, Tokyo]: [Tokyo Tosho], 1980). From *The Annotated Alice*. in two volumes.
3. Trans. [Hiroshi Takayama], [*Shinchu Fushigi no Kunino Arisu*] [New Annotated Alice's Adventures in Wonderland] and [*Shinchu Kagami no Kunino Arisu*] [New Annotated Through the Looking Glass] ([Bunkyo, Tokyo]: [Tokyo Tosho], 1994). From *More Annotated Alice*. in two volumes.
4. Trans. [Hiroshi Takayama], [*Shinchu Fushigi no Kunino Arisu*] [New Annotated Alice's Adventures in Wonderland] and [*Shinchu Kagami no Kunino Arisu*] [New Annotated Through the Looking Glass] ([Bunkyo, Tokyo]: [Tokyo Tosho], 1994). From *More Annotated Alice*. in two volumes.
5. Trans. [Hiroshi Takayama], [*SHochu ArisuKanzen Kettwihan*] [Annotated Alice: Complete definitive edition] ([Bunkyo, Tokyo]: [Akishobo], 2019). From *Annotated Alice: 150th Anniversary*. Not the "definitive" edition.

Juvenile Literature

1. Trans. [Armane Tadashi Yamazaki], [*Shokugo no Bikkuri Kagaku-Kagaku Pazuru de Asobou*] [Wonderful Science After a Meal – Play with Science Puzzle] ([Chiyoda-ku, Tokyo]: [Hakuyousha], 1964). From *Science Puzzlers*
2. Trans. [Nob Yoshigahara and Noritada Nakajima], [*Mahchin Gadona no Pazuru Korekushon*] [Martin Gardner's Puzzle Collection] ([Bunkyo, Tokyo]: [Tokyo Tosho], 1981). From *Perplexing Puzzles and Tantalizing Teasers* and *More Perplexing Puzzles and Tantalizing Teasers*, in one volume.
3. Trans. [Kouichi Kishida and Katsuyuki Ueda], [*Angou de asobu hon*] [The Book for Playing Ciphers] ([Shizensha], 1980). From *Codes Ciphers and Secret Writing*

4. Trans. [Tsutomu Kaneko], [*Uchuu no Pazuru*] [Space Puzzles] ([Chiyoda-ku, Tokyo]: [Hakuyousha], 1972). From *Space Puzzles*

5. Trans. [Jin Akiyama], [*Gadona no Omoshiro Kagaku Jikken*] [Gardner's Fun Science Experiments] ([Tokyo]: [Tokai Daigaku Shuppankai], 1997). From *Entertaining Science Experiments with Everyday Objects*

6. Trans. [Nob Yoshigahara and Noritada Nakajima], [*Learn the Principles of Science Magic*] ([Kodansha] / Bluebacks, 2001). From *Science Magic*

Magic

1. Trans. [Shigeo Araki], [*Macchi no Sokuseki Majutsu*] [Instant Magic with Matches] ([Setagaya-ku, Tokyo]: [Chikara Shobo], 1958). From *Match-ic*

2. Trans. [Shi-Ree], [*Martin Gardner's All Magic*] ([Tokyo Society], 1999). From *Martin Gardner Presents*

3. Trans. [Hitoshi Sato], [*Martin Gardner's All Magic*] (Tokyo: [Hall Publ], 1999). From *Martin Gardner Presents*, part 1

4. Trans. [Hitoshi Sato], [*Martin Gardner's All Magic Continued*] (Tokyo: [Hall Publ], 2002). From *Martin Gardner Presents*, part 2

B.18   Korean

Recreational Mathematics

1. Trans. Chung Ho Lee, [*Aha!: Paro Kugoya*] [Aha! That's It] (Seoul: Sa Key Jul, 1990, 2003). From *Aha!*

2. Trans. Chung Ho Lee, [*Iyagi paradoksu*] [Story of paradox] (Seoul: Sa Key Jul, 199,2003). From *La Magie des Paradoxes* not the later *Aha! Gotcha*

3. Trans. Jink Won Kim, [*Mathematics-Science*] (Seoul: Bonus, 2014). From *Riddles of the Sphinx*

4. Trans. Jink Won Kim, [Mathematician in Wonderland] (Seoul: Purun Media, 2000). From *The Universe in a Handkerchief*

Science

1. Trans. Pak Chin-hui, [*Paengmanin ui sangdaesong iron*] (Seoul: Ungjin Munhwa, 1991*). From *Relativity for the Million*
2. Trans. Kwahak Sedae [group name], [*Mat'in Gaduno ui yang-sonjabi chayon segye*] [Ambidextrous World of Martin Gardner] (Seoul: Kach'i, 1993). From *New Ambidextrous Universe*

Philosophy and Theology

1. Trans. [Yoon-Jae Kang], [*Adam kwa ibu egenun paekkop i issos-sulkka*] (Seoul: Bada Books, 2002). From *Did Adam and Eve have Navels?*

Carrolliana

1. Trans. [Chong-min Choi], [*Koul Nara ui Aellisu*] [Through the Looking-Glass and What Alice Found There] (Seoul: Nara Sarang, 1992). From *The Annotated Alice*
2. Trans. [Inja Choi], *Alice* [Alice in a strange world and Alice in the mirror world] (Seoul: Daehan, 2005). From *The Annotated Alice: The Definitive Edition*

B.19   Latvian

Science

1. Trans. A. Okmanis, *Relativitātes Teorija Visiem* (Rīgā: Izde-vniecība Liesma, 1969). From *Relativity for the Millions*

B.20   Malayalam

Juvenile Literature

1. Trans. Resy George, [*Science Mischiefs*] (Sivakasi: Kerala Sastra Sahithya Parishad, 1988). From *Science Puzzlers*

B.21   Norwegian

Science

1. Trans. Jorgen Randers, *Relativitet for Millioner* (Oslo: J. W. Cappelens Forlag, 1964). From *Relativity for the Millions*

Juvenile Literature

1. Trans. [None given], *Morsomme Forsøk* (Oslo: N. W. Damm, 1963). From *Science Puzzlers*, "Quiz IV" on cover

B.22   Persian

Science

1. Trans. Mahmood Mosaheb, [*Nesbyat baray ha'mekan*] (Tehran: B. T. N. K., 1968). From *Relativity for the Million*

B.23   Polish

Recreational Mathematics

1. Trans. Tomasz Żak, *Moje Najlepsze Zagadki Matematyczne i Logiczne* (Wrocław: Oficyna Wydawnicza Quadrivium, 1998). From *My Best Mathematical and Logic Puzzles*
2. Trans. Tomasz Gronek, *Moje Ulubione Zagadki Matematyczne i Logiczne* (Wrocław: Zysk i S-ka Wydawnictwo, 2018). From *My Best Mathematical and Logic Puzzles*
3. Trans. Paweł Strzelecki, *Ostatnie Rozrywki* (Warszawa: Prószyński i S-ka, 2004). From *The Last Recreations*
4. Trans. Wiktor Bartol, *Wszechświat w Chusteczce* (Warszawa: Prószyński i S-ka, 1999). From *The Universe in a Handkerchief*

Science

1. Trans. Zbigniew Majewski, *Zwierciadlany Wszechświat* (Warszawa: Państwowe Wydawnictwo Naukowe, 1969). From *The Ambidextrous Universe*

2. Trans. Marta Appelt and Dorota Kozińska, *Wielkie Eseje w Nauce* (Warszawa: Prószyński i S-ka, 1998*). From *Great Essays in Science*

Fringe Science

1. Trans. Bronisław Krzyżanowski and Włodzimierz Zonn, *Pseudonauka I Pseudouczeni* (Warszawa: Państwowe Wydawnictwo Naukowe, 1966). From *Fads and Fallacies in the Name of Science*
2. Trans. Piotr Sitarski, *Ekscentryczne Teorie, Oszustwa i Maniactwa Naukowe* (Łódź: Pandora, 1994*). From *Science: Good, Bad, and Bogus.*

   • *Einstein i Parapsychogia* (Łódź: Pandora, 1995*).

3. Trans. Grzegorz Kołodiejczyk, *Jeszcze o New Age: Notatki o Pograniczu Nauki* (Łódź: Pandora, 1995*). From *The New Age*
4. Trans. Grzegorz Kołodiejczyk, *Kamienna Twarz na Marsie i Inne Sensacje New Age* (Łódź: Pandora, 1996*). From *New Age*

Philosophy of Science

1. Trans. Artur Koterski, *Wprowadzenie do Filozofii Nauki* (Warszawa: Fundacja Aletheia, 2000*). Edited. From *Philosophical Foundations of Physics*, Rudolf Carnap

B.24   Portuguese

Recreational Mathematics

1. Trans. [Not given], *Matemática, Magia e Mistério* (Lisboa: Gradiva, 1991*). From *Mathematics, Magic and Mystery*
2. Trans. Bruno Mazza, *Divertimentos Matemáticos: Paradoxos e Jogos Papel* (São Paulo: Ibrasa, 1961, 1967*). From *Mathematical Puzzles and Diversions*
3. Trans. Ana Cristina and Reis e Cunha, *Ah, Descobri: Jogos e Diversões Matemáticas* (Lisboa: Gradiva, 1990). From *Aha! Insight*

   • *Ah, Descobri: Jogos e Diversões Matemáticas* (RBA, 2008).

4. Trans. Jorge Lima, *Ah, Apanhei-te!: Paradoxos de Pensar e Chorar por Mais...* (Lisboa: Gradiva, 1993). From *Aha! Gotcha*

   - *Ah, Apanhei-te!: Paradoxos de Pensar e Chorar por Mais...* (RBA , 2008).

5. Trans. [Not given], *O Festival Mágico da Matemática* (Lisboa: Gradiva, 1994). From *Mathematical Magic Show*
6. Trans. Maria Alice Gomes da Costa, *Rodas, Vida e Outras Diversões Matemáticas* (Lisboa: Gradiva, 1992). From *Wheels, Life and Other Mathematical Amusements*
7. Trans. Jorge Nuno Silva, *As Últimas Recreações* (Lisboa: Gradiva, 2002). From *The Last Recreations*
8. Trans. Vera Ribeiro, *Alice no País dos Enigmas* (Rio de Janeiro: Jorge Zahar, 2000). Gardner introduction. From *Alice in Puzzle-Land* by Raymond Smullyan

Fringe Science

1. Trans. Jorge Régo Freitas, *Manias e Crendices em Nome da Ciência* (São Paulo: Ibrasa, 1960). From *Fads and Fallacies*
2. Trans. Beatriz Sidou, *O Umbigo de Adão: Sobre as Maiores Fraudes da Ciência* (Rio de Janeiro: Ediouro, 2002*). From *Did Adam and Eve Have Navels?*

Carrolliana

1. Trans. Maria Luiza X. de A. Borges, *Alice Edição Comentada: Aventuras de Alice no País das Maravilhas & Através do Espelho* (Rio de Janiero: Jorge Zahar, 2002). From *The Annotated Alice: The Definitive Edition*

B.25   Romanian

Recreational Mathematics

1. Trans. R. Theodorescu, *Amuzamente Matematice* (Bucuresti: Editura Ştiinţifică, 1968). From the first and second books of *Mathematical Puzzles and Diversions*
2. Trans. Alex Butucelea, *Alte Amuzamente Matematice* (Bucuresti: Editura Ştiinţifică, 1970). From *New Mathematical Diversions*

B.26 Russian

Recreational Mathematics

1. Trans. V. S. Bermana, *Matematičeskie Čudesa i Taĭny* [Mathematical Wonders and Secrets] (Moskva: Nauka, 1964, 1967*, 1977*. 1978*,1986*). From *Mathematics, Magic and Mystery*, abridged.
2. Trans. Yuli A. Danilov, *Matematičeskie Golovolomki i Razvlecheniya* [Mathematical Brainbusters and Recreations] (Moskva: Mir, 1971, 1994). From a collection of *Scientific American* columns.

   • *Matematičeskie Golovolomki i Razvlecheniya* [Mathematical Brainbusters and Recreations] (Moskva: Mir, 1999).

   Revised and enlarged.

   • *Matematičeskie Golovolomki i Razvlecheniya* [Mathematical Brainbusters and Recreations] (Moskva: AST/Zebra E, 2010).

3. Trans. Yuli A. Danilov, *Matematičeskie Dosugi* [Mathematical Leisure] (Moskva: Mir, 1972, 2000). From another collection of *Scientific American* columns.
4. Trans. Yuli A. Danilov, *Matematičeskie Novelly* [Mathematical Stories] (Moskva: Mir, 1974, 2000). From a further collection of *Scientific American* columns.
5. Trans. A. V.Bankrashkova, *Novie Matematičeskie Razvlečeniya* [New Mathematical Entertainments] (Moskva: Astrel, 2009). From *New Mathematical Diversions*
6. Trans. M. L. Kulneva, *Neskuchne Matematika: Kaleydoskop Golovolomok* [A Kaleidoscope of Puzzles] (Moskva: Astrel, 2008). From *Mathematical Carnival*
7. Trans. M. L. Kulneva, *1000 Razvivayushshie Golovolomok, Matematiceskie Zagadok i Rebusov Dlya Detey i Vzroslie* [1000 developing puzzles, mathematical puzzles and riddles for children and adults] (Moskva: Astrel, 2010). From *Mathematical Magic Show*
8. Trans. Yuli A. Danilov, *Est' Ideya!* [There is an idea!] (Moskva: Mir, 1982). From *Aha! Insight*
9. Trans. unknown, *Luchshie Matematisčeskie Igry i Golovolomki* [The Best Math Games and Puzzles] (Moskva: Astrel, 2009). From *Mathematical Circus*
10. Trans. Yuli A. Danilova, *A Nu-ka, Dogadaĭsya!* [Come on, guess!] (Moskva: Mir, 1984). From *Aha! Gotcha*

11. Trans. I. E. Zino, *Krestiki-Noliki* [Crosses and Zeroes (Tic-Tac-Toe)] (Moskva: Mir, 1988). From *Wheels, Life and Other Mathematical Amusements*

12. Trans. Yuli A. Danilov, *Puteshestvie vo Vremeni* [Time Travel] (Moskva: Mir, 1990). From *Time Travel and Other Mathematical Bewilderments*

13. Trans. Yuli A. Danilov, *Ot Mozaik Penrouza k Nadežnym Šifram* [From Penrose Tiles to Trapdoor Ciphers] (Moskva: Mir, 1993). From *Penrose Tiles to Trapdoor Ciphers*

14. Trans. Yuli A. Danilov, *Ot Mozaik Penrouza k Nadežnym Šifram [From Penrose Tiles to Trapdoor Ciphers]* (Moskva: Mir, 1993). From *Penrose Tiles to Trapdoor Ciphers*

15. Trans. unknown, *Zagadki Sfinksa i drugie matematicheskie golovolomki* (Moskva: Lenand, 2015). From *Riddles of the Sphinx*

16. Trans. Y. N. Sudareva, *520 (Pyat'sot Dvadtsat') Golovolomok* [520 Puzzles] (Moskva: Mir, 1975). Edited From *536 Puzzles and Curious Problems* by H. E. Dudeney

17. Trans. Yuli A. Danilov, *Alisa v Strane Smekalki* (Moskva: Mir, 1987). Gardner introduction. From *Alice in Puzzle-Land* by Raymond Smullyan

Science

1. Trans. Yu. V. Konobeeva, B. A. Pavlinchuka, N. S. Rabotnova, and V. V. Filippova, *Etot Pravyǐ, Levyǐ Mir* [This Right, Left World] (Moskva: Mir, 1967). From *The Ambidextrous Universe*

2. Trans. Yu. V. Konobeeva, B. A. Pavlinchuka, N. S. Rabotnova, and V. V. Filippova, *Etot Pravyǐ, Levyǐ Mir* [This Right, Left World] (Moskva: LibroKom [KomKniga], 2009*). From *The Ambidextrous Universe*

3. Trans. V. I. Man'ko, K. V. Karadzheva and F. E. Chukreeva, *Teoriya Otnositel'nosti dlya Millionov* (Moskva: Atomizdat, 1965, 1979*). From *Relativity for the Million*

4. Trans. V. I. Man'ko, K. V. Karadzheva and F. E. Chukreeva, *Teoriya Otnositel'nosti dlya Millionov* (Moskva: LibroKom [KomKniga], 2008*, 2010*). From *Relativity for the Million*

Carrolliana

1. Trans. Nina M. Demourova, *Priklyucheniya Alisy v Strane Chudes, Skvoz' Zerkalo i Chto Tam Uvidela Alisa, ili Alisa v Zazerkal'e* (Moskva: Nauka, 1978, 1991). From *Annotated Alice*, with a new preface and an essay by G. K. Chesterton; the 1991 edition also has an afterword by the translator and a bibliography by A. M. Roushaylo.
2. Trans. Nina M. Demourova, *Priklyucheniya Alisy v Strane Chudes* (Moskva: Astrel Publ, 2003). This might be an abridgement of *The Annotated Alice.*
3. Trans. Mikhail Matveev, *Annotirovannaya "Ohota na Snarka"* (Moskva: Trimag, 2014). with new illustrations by Vadima Ivaniuk From *The Annotated Snark.*
4. Trans. Nina M. Demourova, *Alisa v strane chudes. Alisa v Zazerkale* (Moskva: AST, 2015). From *The Annotated Alice: The Definitive Edition.*

Fiction

1. *[Fantasia Mathematica]* (Moskva: Mir, 1982). Contains "No-Sided Professor," and "Island of Five Colors."

Juvenile Literature

1. Trans. M. Stoliar and L. Fomin, *Zanilatel'nye Opyty* [Amusing Experiments] (Moskva: Prosveshcheniye, 1979). From *Science Puzzlers*, adapted but still in English(!), with Russian footnotes and a "Vocabulary" at the end to aid the Russian student.
2. Trans. unknown, *Klassieskie Golovolomki* [Classic Puzzles] (Moskva: Astrel, 2007). From *Classic Puzzles*

B.27 Slovak

Science

1. Trans. Michal Ferianc, *Teória Relativity pre Milióny* (Bratislava: Alfa, 1969). From *Relativity for the Million*

B.28   Slovenian

Mathematics

1. Trans. Gitica Jakopin, *Alica v Dezeli Ugank* (Ljubljana: Drzavna Zalozba Slovenije, 1984). Gardner introduction. From *Alice in Puzzle-Land* by Raymond Smullyan

B.29   Spanish

Recreational Mathematics

1. Trans. Cecilia Abatz, *Magia Inteligente* (Buenos Aires: Zugarto, 1992, 1999*). From *Mathematical Magic and Mystery*

   - Trans. Cecilia Abatz and Elivio Gandolfo, *Matematica, Magia y Misterio* (Barcelona: RBA, 2011*).

2. Trans. Julio Flórez, *Juegos Matemáticos* (Madrid: Revista de Occidente, 1961). From *Mathematical Puzzles and Diversions*
3. Trans. Susana Liberti, *Diversiones Matematicas* (Madrid: Selector, 1989*). From *Hexaflexagons and other Mathematical Diversions* [8 chapters]

   - *Juegos Matemáticos* (Mexico: Selector, 1992*, 1999*).
   - *Acertijos Matemáticos* (Buenos Aries: Cosmos, 1999).
   - *Acertijos Matemáticos* (Mexico: Selector, 2001).

4. Trans. Susana Liberti, *Desafíos Mentales* (Madrid: Selector, 1990, 2001*, 2010*). From *Hexaflexagons and other Mathematical Diversions* [Part 2]
5. Trans. Ma. Teresa Góngora, *Rompecabezas Mentales* (Col. Del Valle, Mexico: Selector, 1991). From *Second Book of Mathematical Puzzles and Diversions* [Part 1]
6. Trans. Ma. Teresa Góngora, *Nuevos Rompecabezas Mentales* (Col. Del Valle, Mexico: Selector, 1991). From *Second Book of Mathematical Puzzles and Diversions* [Part 2]
7. Trans. Luis Bou Garcia, *Nuevos Pasatiempos Matemáticos* (Madrid: Alianza Editorial, 1972, 1997*, 1980, 1982*, 1994*, 1996, 2018). From *New Mathematical Diversions*
8. Trans. Gonzalo del Puerto y Gil, *El Ahorcamiento Inesperado y Otros Entretenimientos Matemáticos* (Madrid: Alianza Editorial, 1991). From *The Unexpected Hanging*

9. Trans. Rafael Millán and Fernando González, *Comunicación Extraterrestre y Otros Pasatiempos Matemáticos* (Madrid: Catedra, 1986). From *Sixth Book of Mathematical Games*

10. Trans. Andrés Muñoz Machado, *Carnaval Matemático* (Madrid: Alianza Editorial, 1980, 1995*, 2018). From *Mathematical Carnival*

11. Trans. [Not known], *Miscelánea Matemática* (Barcelona: Salvat, 1986*, 1993*). From *Mathematical Carnival*

12. Trans. Luis Bou, *Festival Mágico-Matemático* (Madrid: Alianza Editorial, 1984, 1994, 2018*). From *Mathematical Magic Show*

13. Trans. Luis Bou, *¡Ajá! Inspiración ¡Ajá!* (Barcelona: Editorial Labor, 1981, 1983*, 1985*). From *Aha! Insight*

   - *¡Ajá! Inspiración* (Barcelona: RBA, 2007*, 2009*).

14. Trans. Luis Bou, *Circo Matemático* (Madrid: Alianza Editorial, 1983, 1995, 2018). From *Mathematical Circus*

15. Trans. Luis Bou, *¡Ajá! Paradojas: Paradojas que hacen pensar* (Barcelona: Editorial Labor, 1983, 1986*, 1989*, 1994). From *Aha! Gotcha*

   - *¡Ajá! Paradojas que hacen pensar* (Barcelona: RBA Publicaciones, 2007*, 2009, 2013*).

16. Trans. Luis Bou Garcia, *Ruedas, Vida y Otras Diversiones Matemáticas* (Barcelona: Editorial Labor, 1985, 1988). From *Wheels, Life, and Other Mathematical Amusements*

   - *Ruedas, Vida y Otras Diversiones Matemáticas* (Barcelona: RBA, 2008).

17. Trans. Magarita Mizraji, *Juegos y Enigmas de Otros Mundos* (Barcelona: Gedisa, 1987, 1993*, 2000*, 2008*). From *Puzzles from Other Worlds*

18. Trans. Daniel Zadunaisky, *Juegos: Los Mágicos Números del Doctor Matrix* (Barcelona: Gedisa, 1986, 1987*, 2000*, 2011*). From *The Magic Numbers of Dr. Matrix*

19. Trans. Luis Bou Garcia, *Rosquillas Anudadas y Otras Amenidades Matemáticas* (Barcelona: Editorial Labor, 1987). From *Knotted Doughnuts and Other Mathematical Entertainments*

   - *Rosquillas Anudadas y Otras Amenidades Matemáticas* (Barcelona: RBA, 2008, 2010).

20. Trans. Luis Bou Garcia, *Viajes por el Tiempo y Otras Perplejidades Matemáticas* (Barcelona: Editorial Labor, 1988, 1994). From *Time Travel and Other Mathematical Bewilderments*

- *Viajes por el Tiempo y Otras Perplejidades Matemáticas* (Barcelona: RBA, 2007*, 2010).

21. Trans. Luis Bou Garcia, *Mosaicos de Penrose y Escotillas Cifradas* (Barcelona: Editorial Labor, 1989, 1990). From *Penrose Tiles to Trapdoor Ciphers*

   - *Mosaicos de Penrose y Escotillas Cifradas* (Barcelona: RBA, 2008).

22. Trans. Luis Bou Garcia, *Huevos Nudos y Otras Mistificaciones Matemáticas* (Barcelona: Editorial Gedisa, 1996, 2002*, 2010*). From *The Last Recreations*, Volume I

23. Trans. Luis Bou Garcia, *Damas Parábolas y Otras Mistificaciones Matemáticas* (Barcelona: Editorial Gedisa, 1996, 2002*). From *The Last Recreations*, Volume II
   The 2021* reprint is *Damas Parábolas y Mas Mistificaciones*

24. Trans. Luis Bou Garcia, *Las Úlitimas Recreaciones* (Barcelona: Editorial Gedisa, 2002*). From *The Last Recreations*, combined edition

25. Trans. Maria del Pilar Carril Villareal, *Cálculo Diferencial e Integral* (Mexico: McGraw-Hill / Interamericana Editores, 2012). From *Calculus Made Easy* with Silvanus P Thomas

26. Trans. Mirta Rosenberg, *Los Acertijos de Sam Loyd* (Barcelona: Granica, 1988*). Edited. From *Mathematical Puzzles of Sam Loyd, Volume 1*

   - *Los Acertijos de Sam Loyd* (Madrid: Zugarto, 1992*).
   - *Los Acertijos de Sam Loyd* (Buenos Aires: Editorial Juegos and Co., 1999*). (Colección de Mente Acertijos)
   - *Los Acertijos de Sam Loyd* (Barcelona: RBA, 2009*, 2011*). (Colección de Mente Acertijos)

27. Trans. Mirta Rosenberg, *Nuevos Acertijos de Sam Loyd* (Barcelona: Granica, 1989*). Edited. From *Mathematical Puzzles of Sam Loyd, Volume 2*

   - *Nuevos Acertijos de Sam Loyd* (Madrid: Zugarto, 1996*).
   - *Nuevos Acertijos de Sam Loyd* (Buenos Aires: Editorial Juegos and Co., 1999*). (Colección de Mente Acertijos)

28. Trans. Montserrat Millán, *Alicia en el País de las Adivinanzas* (Madrid: Madrid Cátedra, 1986*, 1989*). Gardner introduction. From *Alice in Puzzle-Land* by Raymond Smullyan

   - *Alicia en el País de las Adivinanzas* (Madrid: Grupo Anaya, 2004).

Science

1. Trans. Eli de Gortari, *Maquinas Logicas y Diagramas* (Mexico: Editorial Grijalbo, 1973). From *Logic Machines and Diagrams*
2. Trans. Luis Bou, *Máquinas y Diagramas Lógicos* (Madrid: Alianza Editorial, 1985). From *Logic Machines and Diagrams*, Second Edition
3. Trans. Alonso Ruvalcaba, *Los Grandes Ensayos de la Ciencia* (Mexico: Nueva Imagen, 1998*). From *Great Essays in Science*
4. Trans. Jordi Vilá, *La Explosion de la Relatividad* (Barcelona: Biblioteca Científica Salvat, 1986, 1994). From *The Relativity Explosion*
5. Trans. Fernando García Vela, *Izquierda y Derecha en el Cosmos* (Madrid: Alianza Editorial, 1966). From *The Ambidextrous Universe*
6. *Izquierda y Derecha en el Cosmos* (Barcelona: Biblioteca Científica Salvat, 1973*, 1985). From *The Ambidextrous Universe*
7. Trans. Joan Tarrés Freixenet, *El Universo Ambidiestro: Simetrías y Asimetrías en el Cosmos* (Barcelona: Editorial Labor, 1993). From *The Ambidextrous Universe*

   - *El [Nuevo] Universo Ambidiestro* (Barcelona: RBA Coleccionables / Biblioteca de Divulgación Científica, 1994*). in two volumes.

8. Trans. Diorki, *El Escarabajo Sagrado* (Barcelona: Biblioteca Científica Salvat, 1986, 1995). From *The Sacred Beetle*, in two volumes.
9. Trans. Joseé Javier García Sánz, *La Mente Nueva del Emperador* (Mexico: CNCTFCE, 1996). Gardner introduction. From *The Emperor's New Mind*, by Roger Penrose

   - *La Mente Nueva del Emperador* (Mexico: Fondo de Cultura Econmica, 2002).
   - *La Nueva Mente del Emperador* (Debolsillo, 2006).

Fringe Science

1. Trans. Natividad Sánchez Sáinz-Trápaga, *La Cienca: Lo bueno, lo malo y lo falso* (Madrid: Alianza Editorial, 1988). From *Science: Good, Bad and Bogus*
2. Trans. Juan Pedro Campos Gómez, *La Nueva Era: Notas de un observador de lo marginal* (Madrid: Alianza, 1990*, 2006*, 2007*). From *The New Age*

3. Trans. Jordi Fibla, *Extravagancias y Disparates* [Extravagances and Silly Things] (Barcelona: Alcor, 1988*). From *On the Wild Side*

  • Trans. Jordi Fibla, *Extravagancias y Disparates* [Extravagances and Silly Things] (Barcelona: Martinez Roca, 1993*).

4. Trans. Juan Manuel Ibeas, *¿Tenían Ombligo Adán y Eva?* (Madrid: Debate, 2001). From *Did Adam and Eve Have Navels?*

  • Trans. Juan Manuel Ibeas, *¿Tenían Ombligo Adán y Eva?* (Barcelona: Debolsillo, 2003).

Philosophy of Science

1. Trans. Néster Míguez, *Orden y Sorpresa* (Madrid: Alianza Editorial, 1987*). From *Order & Surprise*
2. Trans. Marco Aurelio Galmarini, *El Ordenador Como Científico y Otros Ensayos Sobre Fantasía y Ciencia* (Barcelona: Ediciones Paidos, 1992). From *Gardner's Whys and Wherefores* [Part 1]
3. Trans. Marco Aurelio Galmarini, *Crónicas Marcianas y Otros Ensayos Sobre Fantasía y Ciencia* (Barcelona: Ediciones Paidos, 1992). From *Gardner's Whys and Wherefores* [Part 2]
4. Trans. Néstor Miguens, *Fundamentación Lógica de la Física* (Buenos Aires: Editorial Sudamericana, 1969). From *Philosophical Foundations of Physics*, Rudolph Carnap

Philosophy and Theology

1. Trans. Josep María Llosa, *Los Porqués de un Escriba Filósofo* (Barcelona: Tusquets Editores, 1989, 2001*). From *The Whys of a Philosophical Scrivener*
2. Trans. Pilar Tutor, et al., *Urantia: Revelación Divina o Negocio Editorial* [Urantia: Divine Revelation or Publishing Business] (SA: Susaeta, 1997*). From *Urantia: The Great Cult Mystery*

  • *Urantia: Revelación Divina o Negocio Editorial* (Barcelona: Tikal Ediciones, 1997*).

Literature

1. Trans. Gerardo Espinosa, *El Mago de Oz* (Madrid (Barcelona): Ediciones Alfaguara (and Salvat), 1987*, 1991*). Gardner epilogue From *The Wizard of Oz* by L Frank Baum

2. Trans. Gerardo Espinosa, *El Mago de Oz* (Barcelona: El Aleph Editores, 2002*). Gardner introduction. From *The Annotated Wizard of Oz* by M. P. Hearn
3. Trans. Unknown, *El Hombre Que fue Jueves* (Sekotia, 20014*). Gardner annotation. From *The Annotated Man Who Was Thursday*

Carrolliana

1. Trans. Francisco Torres Oliver, *Alicia Anotada: Alicia en el País de las Maravillas & A Través del Espejo* (Madrid: Akal, 1987*, 1998*, 2010*). Gardner annotation. From *The Annotated Alice* by Lewis Carroll
2. Trans. Agustín Gervás, *Alicia para los Pequeños* (Madrid: Ediciones Alfaguara, 1977*). Gardner epilogue From *The Nursery Alice* by Lewis Carroll

Oziana

1. Trans. Geraldo Espinosa, *El mago de Oz* (Barcelona: Salvat Alfaguara, 1987). Uses the epilogue from the Dover edition. From *The Wonderful Wizard of Oz* by L. Frank Baum
2. Trans. Concha Cardeñoso Sáenz de Miera, *El Mago de Oz: Edición Anotada* (Barcelona: El Alpeph Editores, 2002). Gardner introduction. From *The Annotated Wizard of Oz*, edited by M. P. O'Hearn

Juvenile Literature

1. Trans. Mirta Rosenberg, *El Idioma de los Espías* [The Language of the Spies] (Madrid: Zugarto, 1991). From *Codes, Ciphers and Secret Writings*
   - *El Idioma de los Espías* (Buenos Aires: Editorial Juegos and Co., 1991). (Colección de Mente Acertijos)
   - *El Idioma de los Espías* (Buenos Aires: Ediciones de Mente, 2006). (Colección de Mente Acertijos)
   - *Matemáticas Para Todos (y Códigos ultrasecretos)* (Barcelona: RBA, 2011*). From both *Entertaining Mathematical Puzzles* and *Codes, Ciphers and Secret Writings*.

2. Trans. Mirta Rosenberg, *Matemática para Divertirse* (Buenos Aires: Granica, 1988*). From *Entertaining Mathematical Puzzles*

- *Matemática para Divertirse* (Madrid: Zugarto, 1994*).
- *Matemática para Divertirse* (Madrid: Ediciones de Mente, 2006*).
- *Matemática para Divertirse* (Madrid: RBA Libros, 2007*, 2014*).
- *Matemáticas Para Todos (y Códigos ultrasecretos)* (Barcelona: RBA, 2011*). From apparently both *Entertaining Mathematical Puzzles* and *Codes, Ciphers and Secret Writings*
- *Matemáticas para Divertirse* (epublibre, 2013*).

3. Trans. [Not known], *Acertijos Divertidos y Sorpredentes* [Amusing and Surprising Puzzles] (Madrid: Zugarto, 1994*, 2000*). From *Perplexing Puzzles and Tantalizing Brainteasers*

- *Acertijos Divertidos y Sorpredentes* (Buenos Aires: Editorial Juegos and Co., 1999). (Colección de Mente Acertijos)
- *Acertijos Divertidos y Sorpredentes* (Buenos Aires: Editorial Juegos and Co., 2000).
- *Acertijos Divertidos y Sorpredentes* (Buenos Aires: RBA., 2009*).

4. Trans. Elvio E. Gandolfo, *Ingenio para Genios* (Buenos Aires: Editorial Juegos, 1996*). From *Classic Brainteasers*

5. Trans. Elvio E. Gandolfo, *Ciencia Magica* (Buenos Aires: Editorial Juegos and Co., 1998). Colección de Mente Acertijos From *Science Magic*

6. Trans. Cecilia Abatz and Elivio E. Gandolfo, *Matematica, Magia, Misterio* (Barcelona: RBA, 2011*). From *Mathematical Magic and Mystery* with *Science Magic*

Magic

1. Trans. Pablo Basterrechea, *Corta la Baraja* (Madrid: Paginas Libros de Magia, 2018*). From *Cut the Cards*

2. *Matemagia* (Paginas Libros de Magia, 2021*). From *Encyclopedia of Impromptu Magic*, volume 1 of 3

3. *Cartommagia* (Paginas Libros de Magia, 2021*). From *Encyclopedia of Impromptu Magic*, volume 2 of 3

4. *Magia de Cerca* (Paginas Libros de Magia, 2021*). From *Encyclopedia of Impromptu Magic*, volume 3 of 3

5. Trans. Georgina Greco, *La Magia de los Magos* (Mexico: Diana, 1973*). Gardner introduction. From *Magician's Magic* by Paul Curry

Miscellaneous

1. Trans. Fernando Blasco, *Puro Abracadabra* (Madrid: Paginas Libros de Magia, 2020). From *Undiluted Hocus-Pocus*

B.30 Swedish

Recreational Mathematics

1. Trans. Birger Stolpe, *Rolig Matematik* (Stockholm: Natur och Kultur, 1960). From *Mathematical Puzzles and Diversions*
2. Trans. Birger Stolpe, *Rolig Matematik: Andra Samlingen* (Stockholm: Natur och Kultur, 1962). From *The Second Book of Mathematical Puzzles and Diversions*
3. Trans. Birger Stolpe, *Rolig Matematik: Tredje Samlingen* (Stockholm: Natur och Kultur, 1968). From *New Mathematical Diversions*

Science

1. Trans. Harry Bökstedt, *Relativitetsteorin* (Stockholm: Wahlström and Widstrand, 1964). From *Relativity for the Million*
2. Trans. Tor Larsson, *Skapelsens Symmetri* (Stockholm: Rabén and Sjögren, 1966). From *The Ambidextrous Universe*

Fringe Science

1. Trans. Leif Björk, *Vetenskap och Humbug* (Stockholm: Natur Och Kultur, 1955). From *In the Name of Science*

B.31   Telugu

Recreational Mathematics

1. Trans. M. Stoliara and L. Fomin, *Vijñānaśāstra Vinōdālu: Mārṭin Gārḍnar Pustaka Anusaraṇa* (Heyderabad: Viśālāndhra, 1986). From *Entertaining Science Experiments*, adapted by the translators

B.32   Turkish

Recreational Mathematics

1. Trans. Neyran Savaşman, *Dr. Matrix ve Gizemli Sayılar* (İstanbul: Güncel Yayıncılık, 2004). From *The Magic Numbers of Dr. Matrix*
2. Trans. Baris Biçakçi, *Hah, Buldum!* (Ankara: Tubitak, 2004). From *Aha! Insight*
3. Trans. Alagan Sezgintüredi, *Matematikçinin Galasksi Rehberi* (İnstanbul: Aylak Kitap, 2011). From *Mathematical Puzzle Tales*

Fringe Science

1. Trans. Celal Kapkin, *Adem İle Havva'nın Göbek Çukurları Var mıydı?* (İstanbul: Evrim Yayınevi, 2000). From *Did Adam and Eve Have Navels?*

B.33   Vietnamese

Science

1. Trans. Dam Xuan Tao, *Thuyet Tuong Doi Cho Moi Nguoi* [Relativism for Everyone] (Hanoi: National University Press, 2002). From *Relativity for the Million*

Scientific American Topics Index

This index supplements the chapter on *Scientific American* column. It was felt that merely listing the columns would not allow people to discover what was contained in them. This index is included, even though it is inevitable that the choice of indexed keywords is idiosyncratic. Typically every year a column of problems would appear, rather than an essay, and indexing those in a useful way is difficult. While some of these problems are in the Topics Index, a separate listing of all the problems is given in the *Scientific American* column chapter.

Most columns appeared in book form with addenda, citing new results and updates from correspondents. Many of the books have been revised and the addenda have been further updated. While addenda are usually attached to the chapters, sometimes they appeared in an appendix. Further, some addenda grew so large that they were presented as separate chapters in the books; so a single column may correspond to one, two or even three chapters. To avoid confusion it was decided to use this rule:

> *Each index entry refers to a book chapter rather than a column, and includes all addenda to that chapter, even if it appeared in an appendix. The final revision of each book was always used.*

The abbreviations below are used. A reference is made to Colossal only when the reference is to addenda appearing only in *The Colossal Book of Mathematics*

First	*Hexaflexagons and Other Mathematical Diversions*
Second	*Second Book of Mathematical Puzzles and Diversions*
New	*New Mathematical Diversions*
DrMatrix	*Magic Numbers of Dr. Matrix*
Unexpected	*Unexpected Hanging*
Sixth	*Sixth Book of Mathematical Diversions*
Carnival	*Mathematical Carnival*
MagicShow	*Mathematical Magic Show*
Circus	*Mathematical Circus*
Wheels	*Wheels, Life, and Other Amusements*
Knotted	*Knotted Doughnuts and Other Mathematical Entertainments*
Time	*Time Travel and Other Mathematical Bewilderments*
Penrose	*Penrose Tiles and Trapdoor Ciphers*
Fractal	*Fractal Music, Hypercards, and More*
Last	*Last Recreations*
Colossal	*Colossal Book of Mathematics*

Scientific American Name Index

This index also supplements the chapter on *Scientific American* column, and uses the same conventions as the previous appendix. This name index lists all names that appeared in the columns.

Names only appearing in a bibliography are not included; when it is unclear when to apply this criterion, the names were included. The same abbreviations are used as in the previous index.

Solomon Golomb had stated that he was the most referenced person in the *Scientific American* column, so of course this was investigated. He appeared in 37 chapters which exceeds Coxeter in 24 but trails Knuth in 46. Conway appeared in 49, while Berlekamp was in 11 and Guy in 19. Dudeney appeared in 52 while Loyd was in 30.

For philosophers we find Archimedes in 9, Aristotle and Plato in 15 each, but Socrates in only 2. Modern philosophers are led by Leibniz and Russell with 19 each, followed by Peirce with 11. And his favorite authors appear often: Baum (11), Carroll (38), Chesterton (9), Dunsany (8), and Nabakov (13).

Koplowitz, Herb, UNEXPECTED–7
Koptsik, A. A., LAST–16
Kordemsky, Boris A., LAST–7
Korzybski, Alfred, NEW–4
Kotani, Yoichi, KNOTTED–6
Kozielski, Dolores, PENROSE–6
Kraaijenhof, Joh.,
    UNEXPECTED–16
Kraft, Dean, TIME–10
Kraitchik, Maurice, FIRST–2, 14,
    SECOND–3, 17, SIXTH–3,
    MAGICSHOW–14, PENROSE–
    10, FRACTAL–15
Krall, A. R., UNEXPECTED–3
Kramer, Hilton, MAGICSHOW–1,
    FRACTAL–8
Kramer, Michael, KNOTTED–21
Krause, Eugene F., LAST–10
Krauss, Rosalind, FRACTAL–9
Kravitz, Sidney, DRMATRIX–7,
    PENROSE–9, FRACTAL–10
Krichtman, Shira, UNEXPECTED–1
Krieger, Michael M., FRACTAL–20
Krieger, Samuel Isaac, WHEELS–8
Krikorian, Nishan, FRACTAL–11
Krippner, Stanley, TIME–10
Kristol, Irving, KNOTTED–21
Kron, R. Vincent,
    MAGICSHOW–15
Kronecker, Leopold, DRMATRIX–1
Kruger, Bill, KNOTTED–15
Kruskal, Joseph B., SECOND–16,
    MAGICSHOW–15,
    LAST–2, 22
Kruskal, Martin David,
    SECOND–16, MAGICSHOW–
    17, CIRCUS–4, WHEELS–8,
    KNOTTED–13, PENROSE–4,
    16, 19, COLOSSAL–28
Kruskal, William, SECOND–16
Krutar, Rudolph A., WHEELS–8
Krutman, Seymour, DRMATRIX–5
Kugel, Peter, SIXTH–12
Kühl, Martin, SIXTH–3
Kuhn, Thomas S., CIRCUS–16,
    TIME–19

Kulagina, Nina, DRMATRIX–20,
    TIME–10
Kulkosky, Edward, DRMATRIX–20
Kulp, G. W., PENROSE–13
Kumbel, see Piet Hien
Kunii, Daizo, SIXTH–15
Kunzell, Ekkehard,
    UNEXPECTED–20
Kurrah, Thabit ibn,
    MAGICSHOW–12
Kurshan, Robert B.,
    DRMATRIX–19
Kyoto, Harold, COLOSSAL–48
Laaser, William T., FRACTAL–11
Lacey, Oliver L., SECOND–19
Laff, Mark, FRACTAL–1
Laffer, Arthur B., KNOTTED–21
Lafferty, Don, LAST–14
Lafferty, V. C., SECOND–7
Lagrange, Joseph Louis,
    MAGICSHOW–8, CIRCUS–17,
    WHEELS–6, KNOTTED–18
Lake, Robert, LAST–14
Lalande, Edna, SIXTH–6
Lam, Clement W. H., NEW–14,
    CARNIVAL–5
Lamb, Charles, UNEXPECTED–17
Lambert, Henry, FRACTAL–12
Lambert, Johann Heinrich,
    CIRCUS–16
Lamé, Gabriel, CARNIVAL–18
Lamphiear, Donald, FIRST–3
Landau, Edmund, SIXTH–6,
    WHEELS–2
Lander, Leon J., WHEELS–2
Landers, Ann, PENROSE–17
Landry, Stephen, PENROSE–19
Landsberg, P. T., TIME–10
Langer, Joel, LAST–5
Langford, C. Dudley, MAGICSHOW–
    5, 11, TIME–6
Langman, Harry, NEW–12,
    UNEXPECTED–11, 16,
    SIXTH–19, CARNIVAL–5,
    TIME–17
Lanska, Douglas J., FRACTAL–11